GUIDELINES FOR LABORATORY DESIGN: HEALTH AND SAFETY CONSIDERATIONS

GUIDELINES FOR LABORATORY DESIGN: HEALTH AND SAFETY CONSIDERATIONS

LOUIS J. DIBERARDINIS
Polaroid Corporation

JANET BAUM
Payette Associates

MELVIN W. FIRST
Harvard School of Public Health

GARI T. GATWOOD
Harvard University

EDWARD GRODEN
Consultant

ANAND K. SETH
Massachusetts General Hospital

A Wiley-Interscience Publication
JOHN WILEY & SONS
New York • Chichester • Brisbane • Toronto • Singapore

Library of Congress Cataloging in Publication Data:

Guidelines for laboratory design.

   "A Wiley-Interscience publication."
   Bibliography: p.
   Includes index.
   1. Laboratories–Design and construction.
2. Laboratories–Safety measures. I. DiBerardinis,
Louis J., 1947-
TH4652.G85   1987   727′.5   87-6063
ISBN 0-471-89134-7

Printed in the United States of America

10 9 8 7 6 5 4 3

Preface

This book is intended to introduce the reader to the health and safety issues related to the design of new or renovated laboratories.

It is hoped that this volume will assist all those involved in the design of laboratories, including laboratory owners, managers and occupants, architects, engineers, health and safety personnel, and risk managers. The book has been written with this wide audience in mind. The authors represent a diverse group of professionals involved in various aspects of laboratory health and safety. They bring the viewpoints of their professions to the complicated and often neglected issues of health and safety aspects of laboratory design.

The material is presented in a manner that will allow the reader to investigate the broad issues of laboratory health and safety design considerations, to evaluate the issues that need to be addressed in a specific type of laboratory (e.g., analytical, teaching) and to evaluate the important aspects of a particular health and safety concern (e.g., layout or ventilation) over a wide variety of laboratory settings.

The book is divided into five major sections: Chapter 1 deals with the health and safety issues related to the entire laboratory building. Chapter 2 discusses those concerns that are common to all types of laboratories. Chapters 3–18 discuss the special needs of 16 specific types of laboratories as determined by the materials used and operations that occur in them. Chapters 19–21 discuss general topics of laboratory design, including energy conservation and issues dealing with actual construction. The Appendixes provide examples of types of laboratory layouts and specifications for some major safety equipment and conditions in the laboratory.

The origin of this book was a contract between DiBerardinis Associates, Inc., Wellesley, MA and Research and Environmental Health Division of Exxon Corporation for the production of a health and safety

v

design manual for petroleum research operations. The present volume
was developed as a more expanded treatment of laboratory design.

The authors wish to gratefully acknowledge Marjory Magowan for her
efforts on the early manuscript and Kathleen Lawton for her assistance in
the final manuscript preparation.

Contents

Abbreviations   ix

Introduction   1

PART I: COMMON ELEMENTS
OF LABORATORY DESIGN

1. Building Considerations   11
2. Laboratory Considerations   55

PART II: DESIGN GUIDELINES FOR A NUMBER OF
COMMONLY USED LABORATORIES

3. General Chemistry Laboratory   89
4. Analytical Chemistry Laboratory   92
5. High-Toxicity Laboratory   95
6. Pilot Plant (Chemical Engineering Laboratory)   101
7. Physics Laboratory   105
8. Clean Room Laboratory   112
9. Controlled Environment (Hot or Cold) Room   120
10. High-Pressure Laboratory   126
11. Radiation Laboratory   130
12. Biosafety Laboratory   136

13. Clinical Laboratory 145

14. Teaching Laboratory 150

15. Gross Anatomy Laboratory 156

16. Pathology Laboratory 164

17. Team Research Laboratory 170

18. Animal Research Laboratory 175

PART III: ADMINISTRATIVE PROCEDURES

19. Bidding Procedures 189

20. Performance and Final Acceptance Criteria 193

21. Energy Conservation 202

PART IV: APPENDIXES

I. Background on HVAC 215

II. ASHRAE Comfort Standards 222

III. Fans 227

IV. ASHRAE Filtration Guide 230

V. Laboratory Hoods and Other Exhaust Air
Contaminant-Capture Facilities and Equipment 231

VI. Exhaust Air Ducts and Accessories 253

VII. Emergency Showers 258

VIII. Emergency Eyewash Units 260

IX. Excess Flow Check Valves 261

X. Signs 263

XI. Matrix 267

References 274

Index 279

Abbreviations

A	Amperes
AMCA	Air Moving & Conditioning Association
ASHRAE	American Society of Heating, Refrigeration and Air Conditioning Engineers
atm	Atmospheres
Bag in, Bag out	A system of changing contaminated filters that reduces operator contamination
cfm	Cubic feet per minute
Ci	Curie
DEQE	Massachusetts Department of Environmental Quality Engineering
DNA	Deoxyribonucleic acid
EPA	U.S. Environmental Protection Agency
FM	Factory Mutual
fpm	Feet per minute
FRP	Fiberglass-reinforced polyester
ft	Feet
ft-c	Foot candles
ft^3	Cubic feet
GFI	Ground fault circuit interruptor
HVAC	Heating, ventilating, and air conditioning

I.D.	Internal diameter
in.	Inches
in. w.g.	Inches of water gauge
JCAH	Joint Commission of the Accreditation of Hospitals
kg	Kilogram
LD_{50}	Lethal dose that kills 50% of the exposed population, usually in animal studies
mW	Milliwatt
NAFM	National Association of Fan Manufacturers
NAS	National Academy of Science
NBS	National Bureau of Standards
NFPA	National Fire Protection Association
ng	Nanogram
NIOSH	National Institute of Occupational Safety and Health
N.O.	Normally open
NRC	Nuclear Regulatory Commission
NSF	National Science Foundation
nsf	Net square feet
OSHA	Occupationation Safety and Health Administration
P.I.	Principal Investigator
psig	Pounds per square inch gage
PVC	Polyvinylchloride
s	Seconds
SAMA	Scientific Apparatus Manufacturers Association
SMACNA	Sheet Metal and Air Conditioning Contractor's National Association, Inc.
TNT	Trinitrotoluene
UL	Underwriters Laboratory

GUIDELINES FOR LABORATORY DESIGN: HEALTH AND SAFETY CONSIDERATIONS

Introduction

1. NEED FOR THIS MANUAL

The construction of new laboratory buildings and the renovation of old ones require close communication between the laboratory users, project engineers, architects, construction engineers, and safety and health personnel. With a multitude of needs to be addressed, all too often safety and health conditions are overlooked or slighted and laboratories may be built with unanticipated safety and health hazards. It is clear that one of the principal objectives of laboratory design should be to provide a safe place in which scientists, engineers, and their staff can perform their work. To fulfill this objective, all safety and health considerations must be evaluated carefully and protective measures incorporated into the design wherever needed.

Over many years, chemists, physicists, biologists, research engineers, and their technicians and assistants have met with injury and death in their laboratories by fire, explosion, asphyxiation, poisoning, infection, and radiation. Injury and death have also resulted from more common industrial accidents, such as falls, burns, and encounters with broken glassware and falling objects. Emphasis on safety usually begins in well organized high school science classes and continues with increasing intensity and sophistication through colleges and graduate schools for the express purpose of educating scientists to observe safe laboratory procedures while learning and to carry this knowledge and experience into their careers. Often, however, the very laboratories in which they later practice their profession are obsolete or, when modern, fail to incorporate safe design principles. Unless such scientists have had the good fortune to observe well designed laboratories, they may be ill equipped to assist architect-engineers with safety design when new or renovated laboratories are being prepared for their use. Few architect-engineers are specialists in laboratory safety and they usually need and welcome the active participation of the scientists to whom the laboratories will be assigned for this type of assistance.

Because laboratory scientists tend to do their work alone or in very small clusters, the dire effects of serious breeches of good safety and health practices seldom result in numerous casualties and for this reason are poorly reported in the popular and professional news media. This may give the impression that the dangers of laboratory work have been exaggerated and perhaps lull scientists into a false sense of security. Accident statistics, however, confirm that laboratories can become dangerous workplaces. Careful thought for worker safety remains an essential part of the laboratory design process.

This manual is designed to provide in a concise, easy-to-use format the information needed by architects and project engineers to design safe and efficient laboratories. It includes safety considerations that must be addressed to comply with governmental regulations as well as recognized good practice standards. Although the manual emphasizes U.S. regulations, it is expected that application of the safety principles cited here will provide safe and efficient laboratories wherever they may be needed.

2. OBJECTIVES OF THIS MANUAL

The purpose of this manual is to provide reliable design information related to specific health and safety issues that should be considered when planning for new or renovated laboratories. The objective will be approached within the framework of other important factors such as efficiency, economy, energy conservation, and design flexibility. Although precise specifications will be provided in some cases, the general intent is to review the relevant safety and health issues and then to recommend appropriate design action, including, where possible, a range of alternatives. In those cases where there are specific U.S. code requirements, the appropriate section of the code will be referenced. In many cases, consultation between project engineers, laboratory users, and industrial hygiene and safety experts will be required at one or more design stages. These instances will be noted in the text and it is the hope of the authors that a relationship characterized by close cooperation and understanding will develop among these groups as a result of the use of this manual.

The manual seeks to address at the design stage the many issues that have a direct bearing on the occupational health and safety of those who work in science and engineering laboratories. It makes no attempt to address all the building structural service requirements that are normal architectural and engineering design considerations, nor does it intend to define good practice laboratory health and safety programs in operating laboratories. It recognizes that all these matters should have an important influence on design considerations and addresses them solely in that context.

3. HOW TO USE THIS MANUAL

A. Design Preparation

Proper use of this manual starts by determining the specific type(s) of laboratories that the intended users require. Numerous distinct laboratory types are described in considerable detail for the reader's guidance in Part II. The safety and health design recommendations for each are based on the operations that are to be performed, as well as the materials and equipment that will be used. It is recognized that laboratory usage patterns tend to change over time and, therefore, it is prudent to try to provide for unique functions with as much design flexibility as possible. In some cases, a predictable changing pattern of usage may call for what we refer to as a general purpose laboratory. We have, therefore, treated the general purpose laboratory as one of the special laboratory design categories.

It is always important for the project manager to communicate frequently with all laboratory users to keep current with their specific needs. Experience has amply confirmed that there is a steep learning curve whenever laboratory personnel enter into the design phase of their own laboratories, and changing requirements are the norm at the start. Because many safety considerations are specific to certain laboratories, but absent in others, it is extremely important that the typical laboratory chosen for design purposes be identified unequivocally as the one the user needs. When dealing with renovation projects, the original building layout must be evaluated very carefully to determine its compatibility with the needs of the intended occupants as well as with the good practice layouts recommended in this manual. Therefore, it will be essential to review each laboratory design recommendation to investigate its compatibility with the building that has been selected for renovation. This must be done cooperatively with the user, the architect and management because critical compromises are almost inevitable when an existing building is adapted to new uses.

Although treated as a major topic in Part III of the manual, energy conservation strategies are discussed and incorporated through the manual. They include heating, cooling, and ventilation systems that minimize the discharge of uncontaminated air; recommendations for the installation of exhaust air devices, for example, fume hoods, that discharge the least air volumes consistent with safety; and the use of fully modulated HVAC systems that supply tempered air consistent with exhaust requirements, but no more. It is anticipated that all ordinary energy conservation measures associated with the laboratory structure will be familiar to architects and engineers and that they will be incorporated into the design of the building. Therefore, these important energy conservation methods will not be discussed in this manual. Instead, only those conservation techniques closely associated with the functioning of occupational health and safety matters will be covered.

All of the technical matter in this manual is divided among three parts and several appendixes in a manner designed to provide easy access to all the occupational safety and health information needed to complete a specific design assignment. This has been accomplished in two ways. *First,* Chapters 1 and 2 (Part I) contain general technical information that applies to all, or nearly all, laboratory buildings (Chapter 1) and laboratory modules (Chapter 2) regardless of the precise nature of the work that will be conducted in each. The purpose of placing all generally applicable information into two early chapters is to avoid repeating it when each of the distinctive types of laboratories is discussed in the individual laboratory-type chapters contained in Part II. The general technical information contained in Part I is also intended to present to the reader a unified body of design principles that will be instructive as well as easily accessible as a reference source.

Part II contains information on detailed specifications, good practice standards, and cautionary advice pertaining to 16 specific types of commonly constructed laboratories for academic and industrial research and for educational purposes. Some laboratories are intended for general purpose usage, for example, undergraduate chemistry teaching, whereas others are intended for very specific and well defined research activities, for example, work with biological hazards. Where appropriate, graduated levels of usage are recognized in the complexity of the facilities that are described; for example, general chemistry teaching laboratories for high school, college, and graduate school instruction.

Part III contains administrative matters pertaining to bidding procedures, final acceptance inspections, and energy conservation considerations. The appendixes (Part IV) contain commonly consulted specifications, consensus and good practice standards, and institutional bid documents and procedures found to be useful in the practice of the authors.

No attempt has been made to treat every conceivable type of laboratory in a separate chapter because some are highly specialized and have too restricted a range of usage to make it worthwhile (for example, total containment biological safety laboratories), whereas others are offshoots of one or more of the laboratories that are described in detail, and the transference of information will be obvious. Where gaps of coverage remain, it is hoped that the general principles enunciated in Part I, plus the specific information contained in Part III, will provide adequate guidance for those confronted with a need to design and construct unique and innovative types of laboratories.

The *second method* used to coordinate the information in the several technical chapters is the use of an invariant numerical classification system throughout the chapters in Parts I and II whereby identical topics are always listed under the same numerical designation. For example, in every chapter in these two parts, all space requirements and spatial organization information will be found in sections numbered 2 after the chapter number. Therefore, sections numbered 1.2 are in Chapter 1 and are concerned with technical aspects of building layout, whereas sections numbered 2.2 are in Chapter 2 and refer to the general technical aspects of laboratory module layout. Similarly, these numbers have been assigned to the chapters that cover each unique type of laboratory. For easier understanding by the reader, the numerical classification system that is used in each chapter throughout Parts I and II of the manual is summarized below.

B. Manual Organization

This manual is organized with sufficient flexibility to guide the user in the design of a complete new, multistory laboratory building, as well as in the renovation of a single laboratory module. It is arranged in a format that allows the user to start with the building description and then to proceed in a logical sequence to the development of each individual laboratory module. The safety and health considerations that need to be addressed in every laboratory design assignment are explained and illustrated in five broad categories.

(1) Guiding Concepts

This section defines each type of laboratory by (a) the nature of the tasks normally performed there, (b) the special materials and equipment used, and (c) the nature of the requirements that contribute to making this laboratory unique. In some instances, hazardous or specialized materials and equipment that should not be used in a particular laboratory are listed to aid in making certain that the architect, project engineer, and laboratory user all have the same laboratory type under consideration. When the laboratory type first selected has serious contraindications for some of the projected activities, an alternative type that does not have such exclusions should be identified and those design guidelines followed.

(2) Laboratory Arrangement

This section discusses and illustrates the area requirement and spatial organization of each type of laboratory with special regard to egress, equipment and furniture locations, and ventilation requirements. Typical good practice layouts are illustrated and a major effort is directed toward calling attention to those that are clearly undesirable. The location of exhaust hoods, biological safety cabinets, clean benches, and items of similar function are given special attention.

(3) Heating, Ventilating, and Air Conditioning (HVAC)

This section describes the desirable elements of a laboratory HVAC system that is designed for comfort and safety. Wherever unique requirements have been recognized because of the critical nature of the work or equipment, they are given special consideration and definition in a special requirements section. Usually, minimum performance criteria are specified, but it should be recognized that somewhat better performance ought to be provided by the design to allow for inevitable system deterioration while in use. Special attention is given to providing adequate makeup air for exhaust-ventilated facilities and to the pressure relationships between laboratories, offices, and corridors. When construction requirements for laboratory systems differ substantially from those that apply to ordinary HVAC installations, the differences are made explicit and appropriate codes and standards are cited in the text.

(4) Loss Prevention, Industrial Hygiene, and Personal Safety

This section presents checklists of items that must be evaluated for their inclusion during the design stages. They encompass a wide variety of safety devices and safety design options intended to protect workers and property. The important subjects of handling dangerous substances and disposal of laboratory waste, including animal waste and animal carcasses, is included. In many cases these items can be attended to later, but usually only at greater expense.

(5) Special Requirements

This section deals with the unique aspects of each identified laboratory type. Not all of the noted special requirements may be needed exclusively for safety reasons, but their presence in a laboratory may affect overall safety considerations and be important for that reason. This section evaluates their potential impact and presents appropriate safety measures when required.

C. Codes and Standards

Governmental and code requirements that pertain to specific safety items are stated and the sources referenced. In the United States, the major codes and regulations that must be met are those of the Occupational Safety and Health Administration (OSHA, 1978), the National Fire Protection Association (NFPA, 1985), and the Building Officials and Code Administrators, International, Inc. (BOCA, 1985). In addition, there are many local codes, ordinances, and state laws that must be observed. In the absence of specific regulations or code requirements, numerous safety-related topics have been treated with special detail because we are of the opinion that our recommendations will have an important impact on improved safety in areas not now adequately addressed elsewhere. In these instances, considerable pains have been taken to justify the recommendations. Whenever possible, alternative recommendations are made to permit flexibility of design and construction, especially for renovation projects where physical constraints are encountered frequently. Even when no specific recommendations have been made, a checklist of items to be considered is often presented. When additional interpretation of recommendations or further explication of design cautions is considered desirable, it is highly recommended that the project engineer and architect work closely with industrial hygiene and safety professionals in an endeavor to design and build the safest laboratories possible.

D. Information Sources

Applicable federal and state regulations, codes and standards, textbooks, and published articles on the safe design of laboratories have been referenced throughout the book to provide the user with more detailed information. Close communication with industrial hygiene and safety professionals throughout the planning phases has been recommended. In the absence of qualified staff personnel, competent professional guidance may be obtained in several ways.

1. Consultants. The American Industrial Hygiene Association* and the American Society of Safety Engineers† maintain current lists of consultants, which can be obtained on request.

2. Governmental Agencies

    (a) Many state Departments of Occupational Health (or Industrial Hygiene) receive Federal assistance for the express purpose of providing professional help for occupational health and safety needs, and they can be called or visited for advice.

    (b) Regional offices of the National Institute for Occupational Safety and Health (in the U.S. Department of Health and Human Services), and the Occupational Safety and Health Administration (in the U.S. Department of Labor), can also be requested to provide answers to specific health and safety issues and interpretation of Federal regulations.

    (c) Local fire departments usually review large renovations and new building plans with respect to fire regulations.

Note. Because it is expected that this manual will be used by people with diverse technical backgrounds, for example, architects, engineers, laboratory scientists, and nontechnical administrators, many terms and concepts have been defined and explained in a degree of detail that may seem excessive to one or another of these professional groups. Should this occur, we hope that the reader will be patient and understanding of the knowledge limits of others on the design team.

---

* American Industrial Hygiene Association, 475 Wolf Ledges Parkway, Akron, OH 44311; (216)762-7294.
† American Society of Safety Engineers, 1800 E. Oakton Street, Des Plaines, IL 60018-2187; (312)692-4121.

COMMON ELEMENTS OF LABORATORY DESIGN

1

Building Considerations

1.1 GUIDING CONCEPTS

This chapter deals principally with alternative building layouts for the design and construction of new laboratory buildings. The advantages and disadvantages of a variety of alternative building designs are presented, along with the preferred choices. Laboratory requirements based on one or another of the preferred building layouts are discussed. During the useful life of a building, laboratories may be renovated several times. Therefore, as much flexibility as possible has been provided so that the health and safety concepts given here may be applied to renovation of existing buildings as well as to original construction. Facilities undergoing simple renovation need not be substantially revised just to meet the requirements contained in this chapter if no safety hazards are present, but consideration should be given to following as closely as possible the precepts contained in this chapter when substantial modifications are to be made. Because renovations of laboratories may be carried out within building layouts that are less than ideal for the purpose, careful consideration and application of health and safety requirements will be required. Because most safety and health requirements can be applied to many different laboratory and building layouts, it should always be possible to meet essential safety requirements.

1.2 BUILDING LAYOUT

1.2.1 The Building Program

The architect and project engineer, with the assistance of the laboratory users, develop the building program from analysis of data collected on (1) personnel who will occupy the building, (2) the research, teaching, and industrial functions to be housed, and (3) the interrelationships of functions and personnel.

1.2.1.1 Program Requirements

DESIGN GOALS. A building program is a written document that describes the design goals for a building. The building program is prepared by a planning team from within the institution or corporation, often with the assistance of a consultant planner. The program describes for whom and where the building will be constructed, and what the building functions will be for the owners and users. Architects and engineers use a building program to learn for whom they are designing the facility, what spaces and facilities are required, and where functions should be located in relation to each other. A building program may also describe performance criteria for building functions that must be accommodated. Usually it does not describe how all of the specifications are to be achieved: that is the architect's task. The program should not be written with a preconceived formal design philosophy in mind, except as may be required to incorporate health and safety guidelines, because the building program is intended to be used by owner and users to evaluate the planning and architectural design that is ultimately developed.

POPULATION NUMBERS AND PROGRAMS. The planners consult those who will occupy leading positions in the new facility for information on population numbers and functions. Leaders include department heads, principal investigators, administrators, and laboratory managers. If the responsible persons are not yet known, the planners consult the administrators of the organization who will hire the primary staff and establish the programs that will be carried out in the building. When no better information is available, allocations of the major divisions of space can be estimated on the basis of the occupancy patterns of buildings of similar function. It may be expected that information from such sources will be less precise than that obtained directly from future occupants.

To estimate the building population when specific numbers are unobtainable, it is sometimes assumed that each principal investigator (P.I.) employs two or three research assistants, two technicians, and three or four trainees, making the population estimate for a principal investigator's laboratory group eight to ten persons. Staffing in some laboratory types may be rather different. Area allocations per person, based on work category, is another way to estimate building populations. For example, if a principal investigator is assigned 2000 nsf (net square feet) for his program needs, the laboratory population estimate will be ten based on the following area allocations: P.I., 800 nsf; four research assistants, 200 nsf each; two technicians, 100 nsf each; and four trainees, 50 nsf each.

Based on survey information on expected numbers of people and functions, the building program will include a listing of the projected laboratory population by laboratory group or special function. The same population data will be used later for the design of each laboratory and each floor of the building.

ROOM TYPES AND FUNCTIONS. The building program will provide a list of all proposed room types and reflect the nature of the work, equipment, and activities that will take place within them. There are four general categories and functions, not including structural and mechanical spaces: (1) laboratories, (2) laboratory support facilities, (3) offices, and (4) personnel support facilities. There are numerous laboratory types: general chemistry, physics, controlled environment, animal, teaching, and so on. Laboratory types are listed in the Program Outline Checklist shown in Table 1-1 and some are discussed in detail in Part II of this manual. Laboratory room types that may be used for teaching must be clearly designated because many states have special codes governing the construction of teaching facilities. Laboratory support facilities include equipment and storage rooms, instrument rooms, data processing facilities, glassware washing rooms, sterilization facilities, media preparation rooms, sample collection or distribution rooms, machine shops, electronics shops, darkrooms, fluorescent microscopy rooms, electron microscopy suites, chemical or flammable liquid storerooms, and storerooms for radioactive, chemical, or biological hazardous wastes. Office facilities include private offices, group offices and secretarial pools, business offices, records offices, and data processing offices. General support facilities include libraries, conference rooms, seminar rooms, mail rooms, stock rooms, shipping and receiving areas, reception areas, toilets, change rooms, locker and shower rooms, health and first aid offices, lounges, meeting rooms, dining facilities, kitchens, and recreation facilities.

SHARED AND PROPRIETARY FACILITIES. Questions concerning the use of centralized versus proprietary facilities must be answered before it is possible to estimate the number of each of the rooms and laboratories that will appear on the room-type list. The building program addresses the issue of which facilities will be repeated on each laboratory floor, which will be shared by occupants of a department, which will be common to all occupants of the building, and which will be provided outside the building. For example, cold rooms may be provided on each floor, but only one radiation laboratory may be provided for all members of a department. A shipping and receiving dock is an example of a single facility for an entire building. An example of a needed facility that is best located exterior to the laboratory building in a separate structure is a flammable chemical storage facility. Although it is often more economical to build centralized laboratories and laboratory support facilities rather than to duplicate them for each department or laboratory group, the continuing costs of administering centralized services must be taken into account. In the programming phase, the planning team resolves all issues of centralized versus proprietary facilities with the principal occupants and with the owners. After that, the number of each type of room can be estimated.

NET ASSIGNABLE AREA. There are additional issues related to area allocation. The building program should contain guidelines for the architects and engineers on the optimal amount of net assignable area by location for use within the projected total floor area. Figures 1-1 through 1-4, Laboratory Building Configurations, show variations of 40 to 80% net assignable laboratory area on a typical floor in these examples.

The amount of laboratory area available in a building can be increased at a later time by converting nonlaboratory areas, such as offices, stockrooms, and personnel support areas, into laboratories. However, to do this efficiently and at least cost, advance planning is required to provide the needed capacity in the building systems used for heating, ventilating, and air conditioning, electrical services, and piped utilities. The demand and capacity standards for ventilation, cooling, electricity, water, waste drainage, gas, and so on, are far greater for laboratories than for nonlaboratory areas. Normal engineering diversity factors for electrical capacity do not apply for laboratory use. The constant connected load is very high for laboratory equipment, some of which is very rarely turned off. Therefore, the building program should identify the rooms and spaces lacking specific information of this kind, or specify the proportion of nonlaboratory area that may be engineered for conversion to laboratories some time in the future.

RELATIONSHIPS BETWEEN SPACES AND FUNCTIONS. The next building program task is to inform the architects and engineers of the important relationships between the parts of the building that were identified above. The building program does not specify what is on every floor; that

AREA ANALYSIS

		GROSS AREA	NET AREA	MECHANICAL	CIRCULATION	LABORATORY	LAB SUPPORT	OFFICES	% LAB/GROSS	% USEABLE/GROSS
PLAN A	average	11,000	9,605	880	1,760	5,136	0	2,940	46%	73%
	minimum	7,800	6,780	500	1,000	3,880	0	1,900	50%	74%
PLAN B	average	7,260	6,380	400	880	5,500	0	0	75%	75%
	minimum	5,300	4,710	300	500	4,210	0	0	80%	80%

PLAN A

DIMENSIONS

minimum average

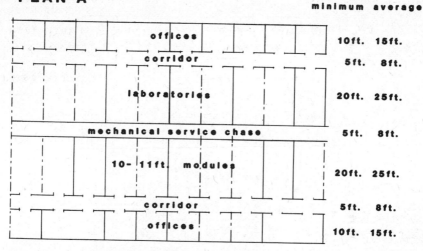

	minimum	average
offices	10ft.	15ft.
corridor	5ft.	8ft.
laboratories	20ft.	25ft.
mechanical service chase	5ft.	8ft.
10- 11ft. modules	20ft.	25ft.
corridor	5ft.	8ft.
offices	10ft.	15ft.

PLAN B

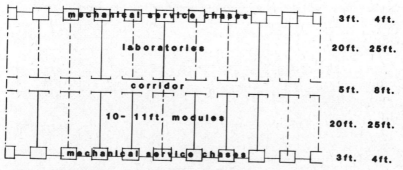
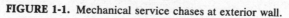

	minimum	average
mechanical service chases	3ft.	4ft.
laboratories	20ft.	25ft.
corridor	5ft.	8ft.
10- 11ft. modules	20ft.	25ft.
mechanical service chases	3ft.	4ft.

FIGURE 1-1. Mechanical service chases at exterior wall.

AREA ANALYSIS

		GROSS AREA	NET AREA	MECHANICAL	CIRCULATION	LABORATORY	LAB SUPPORT	OFFICES	% LAB/GROSS	% USEABLE/GROSS
PLAN C	average	11,000	9,815	960	1,360	5,135	3,320	0	46%	77%
	minimum	8,000	7,120	700	740	3,880	2,500	0	46%	80%
PLAN D	average	7,480	6,700	600	1,240	5,460	0	0	72%	72%
	minimum	5,300	4,750	400	780	3,970	0	0	75%	75%

PLAN C

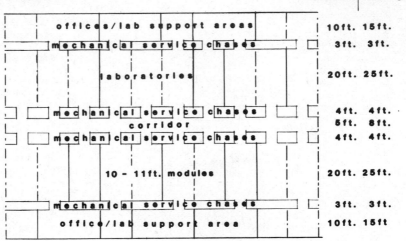

PLAN D

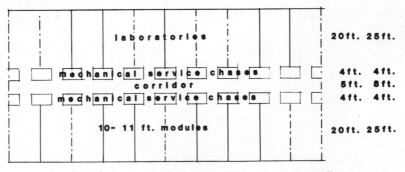

FIGURE 1-2. Mechanical service chases at interior wall.

AREA ANALYSIS

		GROSS AREA	NET AREA	MECHANICAL	CIRCULATION	LABORATORY	LAB SUPPORT	OFFICES	% LAB/GROSS	% USEABLE/GROSS
PLAN E	average	12,320	11,445	220	1,760	6,745	0	2,940	55%	79%
	minimum	7,800	7,220	142	1,000	4,320	0	1,900	55%	80%
PLAN F	average	7,590	6,960	220	1,320	5,640	0	0	74%	74%
	minimum	5,700	5,210	142	1,000	4,210	0	0	74%	74%

PLAN E

DIMENSIONS
minimum average

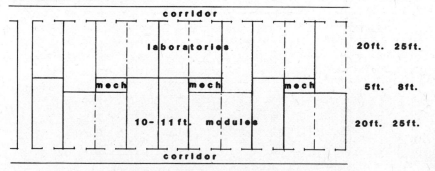

	minimum	average
offices	10ft.	15ft.
corridor	5ft.	8ft.
laboratories	20ft.	25ft.
mech	5ft.	8ft.
10- 11ft. modules	20ft.	25ft.
corridor	5ft.	8ft.
offices	10ft.	15ft.

PLAN F

	minimum	average
corridor		
laboratories	20ft.	25ft.
mech	5ft.	8ft.
10- 11ft. modules	20ft.	25ft.
corridor		

FIGURE 1-3. Central mechanical service chase.

17

AREA ANALYSIS

		GROSS AREA	NET AREA	MECHANICAL	CIRCULATION	LABORATORY	LAB SUPPORT	OFFICES	% LAB/GROSS	% USEABLE/GROSS
PLAN G	average	6,670	5,305	960	935	4,370	0	0	65%	65%
	minimum	4,970	4,020	600	540	3,480	0	0	70%	70%
PLAN H	average	11,115	9,360	980	1,872	4,425	0	3,063	40%	67%
	minimum	7,885	6,625	615	1,080	3,535	0	2,010	45%	70%

PLAN G

DIMENSIONS
minimum average

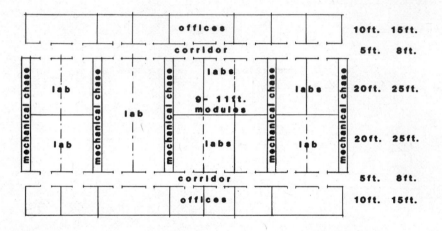

offices — 10ft. 15ft.
corridor — 5ft. 8ft.
20ft. 25ft.
20ft. 25ft.
corridor — 5ft. 8ft.
offices — 10ft. 15ft.

PLAN H

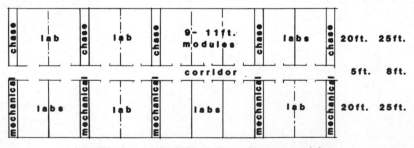

20ft. 25ft.
corridor — 5ft. 8ft.
20ft. 25ft.

FIGURE 1-4. Mechanical chases between modules.

EVALUATION OF SERVICE CHASE AT EXTERIOR WALL

ADVANTAGES:

short horizontal exhaust duct from fume hood
fume hood is at the end of an aisle
major inner lab traffic near corridor egress
equipment and desks at corridor wall

DISADVANTAGES:

pipes and ducts only accessible from laboratory
secondary egress near corridor exit
restricted window area
outswinging doors temporarily obstruct corridor, wide corridors are recommended

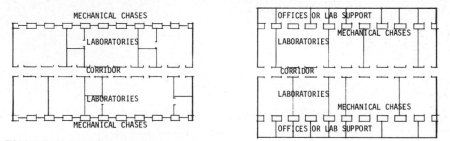

BUILDING CONFIGURATIONS

FIGURE 1-5. Building configurations: Evaluation of service chase at exterior wall.

information is developed and organized in the planning process. A list of questions helpful for establishing important relationships between spaces and functions follows:

1. What is the organizational structure of the institution or corporation for whom the building is being designed? Should room assignments and groupings reflect this hierarchy?
2. Do materials, processes, and waste products contained or produced in one area affect the function or pose a hazard for any other area or function? If the answer is affirmative, what arrangements can be made to reduce or eliminate conflicts?
3. How close to the laboratories they serve do certain laboratory support facilities have to be, and are there critical relationships between them that affect health, safety, or efficiency?
4. How close to laboratories do offices have to be? Must offices be contiguous with laboratories, across the hall, or in a separate wing?

EVALUATION OF SERVICE CHASE AT INTERIOR CORRIDOR

ADVANTAGES:

pipes and ducts accessible from corridor
secondary egress opposite corridor
expansive window area possible on perimeter wall
outswinging door shielded in alcove
equipment and desks at window wall

DISADVANTAGES:

long horizontal exhaust duct from fume hood
persons pass in front of fume hood, causing turbulence

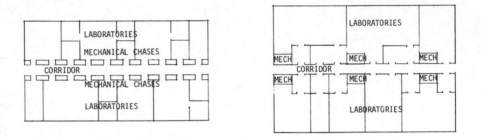

BUILDING CONFIGURATIONS

FIGURE 1-6. Building configurations: Evaluation of service chase at interior wall.

5. Do certain laboratories, or mechanical services to those facilities, need to be isolated from other building functions or services for reasons of health, safety, or as a necessary part of their procedures and equipment operation?

Answers to these questions provide additional information to assist the architects and engineers in the preparation of the building program that precedes the design of the building. The building program outlines performance specifications and states the owner's goals for building function. For the owner, the building program is a document against which the architectural design can be evaluated for adherence to the stated goals and specifications. To further assist the architects and engineers, the laboratory procedures and safety manuals used by the building owners should accompany the building program to alert the designers to particular health and safety concerns of the owners. When such documents have

EVALUATION OF SERVICE CHASE BETWEEN MODULES

ADVANTAGES:

short horizontal exhaust duct from fume hood
fume hood is at the end of an aisle
major inner lab traffic near corridor egress
expansive window area possible on perimeter wall
applicable to conversion of building to laboratory use

DISADVANTAGES:

single module lab has only one wet wall
secondary egress between labs not possible in plans H and I
outswinging doors temporarily obstruct corridor, wide corridors are recommended
pipes and drains have long horizontal runs to peninsula benches
drains require additional venting

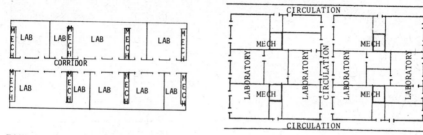

BUILDING CONFIGURATIONS

FIGURE 1-7. Building configurations: Evaluation of service chase between modules.

not yet been prepared, borrowed manuals for similarly engaged facilities
or appropriate sections of this manual may be substituted.

CHECKLIST. The checklist in Table 1-1 provides a skeletal overview of
the building program elements. It is not complete but is intended to serve
as a general review to assure consideration of all major areas.

1.2.1.2 Area Guidelines

The density and intensity of use for each identified type of space can be
estimated by defining the floor area needed for each individual process
and employee, or classification of employees. From this, the overall
amount of assignable space for the entire facility can be summed. The
estimated total building area then becomes the sum of the floor areas for
all known assignable spaces adjusted by a factor that roughly accounts
for the proportion of nonassignable floor area. The building program

EVALUATION OF CENTRAL SERVICE CHASE

ADVANTAGES:

minimum mechanical area, efficient for low buildings
fume hood is at the end of an aisle
short horizontal exhaust duct from fume hood
pipes and ducts for cluster of 4 labs accessible from each
maximum variation in laboratory size possible

DISADVANTAGES:

limited chase area not suitable for multi-storied bldgs.
no window area in laboratory
outswinging doors temporarily obstruct corridor, wide corridors are recommended

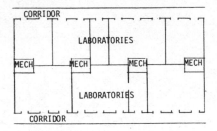 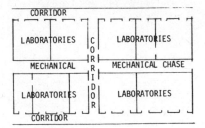

BUILDING CONFIGURATIONS

FIGURE 1-8. Building configurations: Evaluation of central service chase.

TABLE 1-1
Program Outline Checklist

A. Within each laboratory:
   Determine the projected number of full-time-equivalent persons occupying each laboratory and laboratory support area. Designate job classifications where possible.
   Determine the number, types, and area for workstations required by the personnel listed above.

Commonly Encountered Laboratories and Laboratory Support Facilities

Laboratories	Laboratory Support Facilities
General chemistry	Equipment room
Analytical chemistry	Instrument room
High toxicity	Data processing area
Pilot plant	Glass washing and sterilization room
Physics	Media preparation and tissue culture room
Clean room	Sample collection or distribution room

Table 1-1 (*Continued*)

Controlled environment	Machine shop
High pressure	Electronics shop
Biological safety	Darkroom
Clinical	Microscopy room
Animal housing	Electron microscopy suite
Teaching	Chemical or flammable liquid storerooms
Radiation	Radioactive and hazardous waste storeroom
Anatomy/pathology	Balance room
Team research	Shipping and receiving area

B. Within each department or on each floor:

Determine the projected number of full-time-equivalent persons occupying each office or nonlaboratory work area, designating job classifications where possible.

Determine the number, types, and area of workstations required by office and office support personnel.

Commonly Encountered Offices and Office Support Facilities

Offices	Office Support Facilities
Private office	Data processing area
Group office	Library or resource room
Business or records office	Conference room
	Meeting or seminar room
	Mail room
	Stockroom
	Shipping and receiving area

C. Within the building:

Estimate from the totals in A and B above the number of persons who will use each of the following building-wide personnel support facilities.

Estimate the frequency and intensity of use for each building-wide personnel support facility identified from the following list to determine the approximate area required.

Commonly Encountered Building-Wide Personnel Support Facilities

Lounge/meeting rooms
Dining facilities (kitchen, lunchroom)
Toilets, changing rooms, locker rooms, showers
Health or first aid office
Reception area
Recreation/exercise rooms

documents the total building area, space by space, with information on the functional relationships and personnel density of each identified space.

The following are terms planners and architects use to characterize area statistics.

Gross Area. The total building. It includes areas taken up by structure, exterior walls, partitions, and mechanical chases, plus all usable areas.

Net Usable Area. The total area within the exterior walls excluding the area taken up by the structure and partitions. Net usable area includes circulation, maintenance, mechanical, and all assignable areas as well as nonassignable floor areas such as corridors, stairs, elevators, and mechanical chases.

Net Assignable Area. The total area of rooms and spaces on the room-type list.

The following are terms planners and architects use to characterize room types.

Laboratory. A category of net assignable area, not including partitions or vertical chases, in which diverse mechanical services and special ventilation are available.

Laboratory Support Area. Contains the same services as the laboratory area but may or may not conform to the same modular laboratory configuration.

Office Area. A category of net assignable area that does not include partitions or vertical chases. It contains only electrical and conventional office ventilation services.

Personnel Support Areas. Although similar to office areas, may contain additional mechanical services and nonstandard ventilation services to provide for special needs.

If the size of the building to be constructed or renovated has not already been determined by budget or site restrictions, the total area needed can be estimated from the data presented in the building program. As noted, the method used is to estimate the area, or range of areas, that each room type may occupy and then to calculate the sum of all the areas for each room type. This number represents the *estimated total net assignable area.* As shown in Figures 1-1 to 1-4, the percentage of *net usable area* on a typical floor varies from *65 to 80% of the gross area.* However, when the whole building is considered, with areas included for vertical circulation, mechanical and maintenance rooms, and personnel support facilities, gross area usually turns out to be approximately double the estimated *total net assignable area*. The total gross area figure is used by architects and engineers in all phases of spatial and engineering design. Construction, operating, and maintenance costs are also estimated from the total gross area figure. Therefore, if building cost is $150/ft^2, assignable laboratory space will cost $300/ft^2.

1.2.2 Planning

The building program gives the architects and engineers the information they need to plan the building. The planning process includes all the issues to be described in sections 1.2.2.1 through 1.2.2.7. The end product of the planning process is a set of schematic design drawings and specifications for materials. Architects customarily show in schematic design drawings their initial concepts of what the building will look like, including, height, shape, volume, and materials to be used.

Schematic design drawings should also show the size and location of every room listed in the building program by type and by floor or level. The drawings should indicate the distribution of mechanical services throughout the building, how the building will be arranged on the site, and the means of legal ingress and egress. The structural system may be proposed in the schematic design. Preliminary room and area assignments for the proposed occupants of the building may be shown on the schematic design drawings when the final occupants are known.

Owners, occupants, architects, and engineers should communicate intensively during the planning process to exchange information and to resolve problems that were not foreseen during development of the building program.

1.2.2.1 Building Spatial Organization

While the optimal building enclosure configuration is being designed, concepts for the internal organization of the building spaces must be generated. The internal organization comprises six major patterns of spatial definition:

1. Circulation of people and materials (Section 1.2.2.2).
2. Laboratory module (Section 1.2.2.3).
3. Distribution of mechanical equipment and services (Section 1.2.2.4).
4. Structural system (Section 1.2.2.5).

5. Site regulations (Section 1.2.2.6).

6. Building enclosure (Section 1.2.2.7).

These patterns of spatial definition often conflict when not considered together. Although structural and mechanical engineers can design many solutions to fit the building enclosure, a comprehensive concept is needed to fulfill building program requirements. All systems should complement one another to provide a safe and healthful work environment at reasonable cost.

1.2.2.2 Circulation of People and Materials

A major determinant of spatial definition is the circulation of people and materials within the building and around it. The health and safety issues of circulation are primarily concerned with emergency egress and access to the building and its internal parts by emergency personnel such as firemen and police. The major references used for preparing this part of the manual, that is, the Building Officials and Code Administrators International, Inc. (BOCA, 1985), National Fire Protection Association (NFPA, 1985), and Occupational Safety and Health Administration (OSHA, 1978), define and specify all the components of building egress and access that pertain to safety. The specific sections are the following: (1) BOCA Section 600; (2) NFPA Bulletin 101, Life Safety Code, Chapter 5, Means of Egress; Section 13-2, Business Occupancies, Means of Egress Requirements; and Section 14-2, Industrial Occupancies, Means of Egress Requirements; (3) OSHA Health and Safety Standards, 29CFR 1910., Subpart E, Means of Egress.

Special attention should be given to ease of access and emergency evacuation for handicapped persons. The 1982 *Rules and Regulations* of the Architectural Barriers Board of the Commonwealth of Massachusetts, (Mass. 1982) is a model of the codes proposed and adopted by many states. Buildings that comply with these codes facilitate employment of handicapped persons and their integration into the workforce. Architectural barriers should be avoided at main entrances to buildings, doorways, public toilet rooms, elevators, drinking fountains, and public telephones. Suitable door hardware and surrogates for visible signs should be provided for those who are visually impaired.

1.2.2.3 Laboratory Module

A laboratory module is defined as a basic unit of space of a size commonly referred to as a two-person laboratory. Formulation of the internal organization of the laboratory building begins with a decision on the dimensions of the laboratory module. This redirects the planning focus from the large scale of the total facility down to the small scale of a single laboratory.

1.2.2.3.1 *Laboratory Width*

Criteria for the individual laboratory work area have been studied in detail by the Nuffield Foundation, United Kingdom (Nuffield, 1961). Their time/motion efficiency studies led to specifications for optimal dimensions of the standard laboratory aisle. The laboratory aisle is a space, usually flanked by an array of work surfaces, equipment, benches, and utilities, where laboratory personnel spend their workday. The aisle between benches, work surfaces, or equipment should be a minimum of 60 in. so that a person can pass behind another person working at a bench. The maximum clearance should be 72 in. because aisles wider than this tend to get clogged with free standing equipment and other obstructions.

To assure that there is sufficient space for a single-aisle laboratory at each module, the dimension for the thickness of a wall should be added to the clear width of each module. This provides flexibility to rearrange an individual single-module laboratory, or an entire laboratory floor, without reducing the recommended aisle width. Special purpose laboratories, such as controlled-environment laboratories and pilot plants, have space requirements that may not conform to the standard laboratory module.

Existing structures converted to laboratory use may not have structural module dimensions consistent with division into the recommended module dimensions. To overcome this difficulty, adjustments may have to be made on the depth of benches or in the distribution of utilities to the laboratory floor, but under no circumstances should the clear aisle width be reduced below 60 in. Table 1-2 presents laboratory widths for a variety of modular arrangements.

TABLE 1-2
Laboratory Width by Number of Modules in a Building Unit

No. of Modules	1	2	3	4	5	6
Number of parallel rows						
Aisles	1	2	3	4	5	6
Benches	2	4	6	8	10	12
Utility strips	2	4	6	8	10	12
Width of parallel rows						
Aisles at 5'-0"	5'-0"	10'-0"	15'-0"	20'-0"	25'-0"	30'-0"
Benches at 2'-0"	4'-0"	8'-0"	12'-0"	16'-0"	20'-0"	24'-0"
Utilities at 6"	1'-0"	2'-0"	3'-0"	4'-0"	5'-0"	6'-0"
Total constructed width, center to center						
Walls 6" thick	10'-6"	21'-0"	31'-6"	42'-0"	52'-6"	63'-0"
Walls 8" thick	10'-8"	21'-4"	32'-0"	42'-8"	53'-4"	64'-0"

1.2.2.3.2 Laboratory Length

The length of a laboratory module is governed by several variables: the overall width of the building enclosure, the structural span, and the area allotment for a standard module. Laboratory module length is generally 20 to 30 ft for efficient operation of the laboratory. Laboratory length in excess of 30 ft may generate egress problems.

1.2.2.3.3 Laboratory Unit

An assembly of a number of laboratory modules, access corridors, and contiguous accessory spaces into a larger space category becomes a laboratory unit. NFPA 45, "Standard on Fire Protection for Laboratories Using Chemicals" (NFPA 45, 1985), specifies criteria for planning laboratory units. Two important tables in this bulletin are "Maximum Quantities of Flammable and Combustible Liquids in Laboratory Units Outside of Approved Flammable Liquid Storage Rooms," reproduced here as Table 1-3, and "Construction and Fire Protection Requirements for Laboratory Units," reproduced here as Table 1-4. Table 1-3 establishes hazard classifications of laboratory units on the basis of the number of gallons of flammable liquids stored within the laboratory unit per 100 nsf area. The philosophical basis of Table 1-3 is that the amount of flammable liquids allowed within a laboratory unit is directly proportional to its size, up to a defined maximum quantity. Table 1-4 lists minimum fire separation requirements for laboratory units in each hazard classification. From Table 1-4, the maximal or optimal size of a laboratory unit can be estimated. In actual practice, hazard classification of laboratory units is made after the building is occupied, rather than in the planning stage, but consideration must always be given in the planning process to a reasonable subdivision of space to limit the spread of fire, fumes, and other hazards that may threaten life or cause extensive property damage.

1.2.2.4 Distribution of Mechanical Equipment and Services

Mechanical engineers, in consultation with the architect and project engineer, design the distribution of ventilation air, mechanical equipment, and piped utilities. Plans for recommended laboratory layouts and distribution of services are presented in Figures 1-5 to 1-8. These layouts are categorized by the location at each module of the vertical chases that contain risers of piped utilities and ventilation ducts, as follows:

Figure 1-5—Evaluation of service chase at exterior wall.
Figure 1-6—Evaluation of service chase at interior wall.

TABLE 1-3
Maximum Quantities of Flammable and Combustible Liquids in Laboratory Units Outside of Approved Flammable Liquid Storage Rooms

Laboratory Unit Class	Flammable or Combustible Liquid Class	*Excluding* Quantities in Storage Cabinets and Safety Cans[g]			*Including* Quantities in Storage Cabinets and Safety Cans[g]		
		Maximum Quantity[c] Per 100 ft² of Laboratory Unit (gal)	Maximum Quantity[d] Per Laboratory Unit (gal) Unsprinklered	Maximum Quantity[d] Per Laboratory Unit (gal) Sprinklered[f]	Maximum Quantity[c] Per 100 ft² of Laboratory Unit (gal)	Maximum Quantity[d] Per Laboratory Unit (gal) Unsprinklered	Maximum Quantity[d] Per Laboratory Unit (gal) Sprinklered[f]
A[a] (High hazard)	I	10	300	600	20	600	1200
	I, II, and IIIA[e]	20	400	800	40	800	1600
B[b] (Intermediate hazard)	I	5	150	300	10	300	600
	I, II, and IIIA[e]	10	200	400	20	400	800
C[b] (Low hazard)	I	2	75	150	4	150	300
	I, II, and IIIA[e]	4	100	200	8	200	400

[a] Class A Laboratory units shall not be used as instructional laboratory units.

[b] Maximum quantities of flammable and combustible liquids in Class B and Class C instructional laboratory units shall be 50% of those listed in the table.

[c] For maximum container sizes, see Table 7-2, NFPA 45

[d] Regardless of the maximum allowable quantity, the maximum amount in a laboratory unit shall never exceed an amount calculated by using the maximum quantity per 100 ft² of laboratory unit. The area of offices, lavatories, and other contiguous areas of a laboratory unit are to be included when making this calculation.

[e] The maximum quantities of Class I liquids shall not exceed the quantities specified for Class I liquids alone.

[f] Where water may create a serious fire or personnel hazard, a nonwater extinguishing system may be used instead of sprinklers.

[g] For SI units: 1 gal = 3.785 L;100 ft² = 9.3 m².

Source. National Fire Protection Association, Quincy, MA, 1985.

29

TABLE 1-4
Construction and Fire Protection Requirements for Laboratory Units[a]

Laboratory Unit Fire Hazard Class	Area of Laboratory Unit Square Feet	Nonsprinklered Laboratory Units — Construction Type I and II[c]		Nonsprinklered Laboratory Units — Construction Types III, IV, and V[c]		Spinklered Laboratory Units[b] — Any Construction Type[c]	
		Separation from Nonlaboratory Area	Separation from Lab. Units of Equal or Lower Hazard Classification	Separation from Nonlaboratory Areas	Separation from Lab. Units of Equal or Lower Hazard Classification	Separation from Nonlaboratory Areas	Separation from Lab. Units of Equal or Lower Hazard Classification
A	Under 1000	1 h	1 h	2 h	1 h	1 h	NC/LC[c,d]
	1001–2000	1 h	1 h	N/A[d]	N/A	1 h	NC/LC
	2001–5000	2 h	1 h	N/A	N/A	1 h	NC/LC
	5001–10,000	N/A[d]	N/A	N/A	N/A	1 h	NC/LC
	10,001 or more	N/A	N/A	N/A	N/A	N/A	N/A
B	Under 20,000	1 h	NC/LC[c,d]	1 h	1 h	NC/LC[e,g]	NC/LC
	20,000 or more	N/A	N/A	N/A	N/A	N/A	N/A
C	Under 10,000	1 h	NC/LC	1 h	NC/LC	NC/LC[e,g]	NC/LC[e,g]
	10,000 or more	1 h	NC/LC	1 h	1 h	NC/LC	NC/LC

[a] Where a laboratory work area or unit contains an explosion hazard, appropriate protection shall be provided for adjoining laboratory units and nonlaboratory areas, as specified in Chapter 5, NFPA 45

[b] In laboratory units where water may create a serious fire or personnel hazard, a nonwater extinguishing system may be substituted for sprinklers.

[c] See Appendix B-4, NFPA 45

[d] N/A, not allowed: NC/LC, noncombustible/limited combustible construction. (See Appendix B-4, NFPA 45)

[e] May be 1/2 hour fire-rated combustible construction.

[f] Existing combustible construction is acceptable.

[g] Laboratory units in educational occupancies shall be separated from nonlaboratory areas by 1 h construction. For S1 units: 1 ft² = 0.929 m².

Source. National Fire Protection Association, Quincy, MA, 1985.

Figure 1-7—Evaluation of service chase between modules.

Figure 1-8—Evaluation of central service chase.

Laboratory buildings require a great amount of energy to supply, condition, and exhaust ventilation air. Vertical distribution of ducts is the efficient design approach for exhaust systems. Continuity of vertical chases from floor to floor and through the entire building to the equipment on the roof is necessary for this organization. Therefore, functions that require a large undivided floor area (such as an auditorium or library) cannot be located on top of laboratory floors or sandwiched between laboratory floors without loss of usable floor area and possible serious interference with intended functions. Vertical chases at each module allow a laboratory hood or other special exhaust system to be installed there at any time. The same flexibility is available for piped utilities to each module. The floor area occupied by vertical chases ranges between 1 and 10% of the net usable area on a typical laboratory floor. Building supply air is usually a combination of vertical and horizontal systems. Following is a discussion of the building layouts shown in Figures 1-5 through 1-8.

There are advantages and disadvantages associated with placement of vertical service chases at the rear of the laboratory module. An important advantage is that the laboratory hood will be located away from the door and adjacent to the vertical chase, favoring a short, energy-efficient connection to the main duct. Desks, which should represent the area of lowest potential hazard in the laboratory, can then be placed at the corridor wall, near the primary egress. This arrangement permits pipes and ducts in the vertical chases to be maintained and serviced from a service corridor outside the laboratory modules. The service corridor itself provides a second egress from each laboratory module and becomes an emergency egress if it is constructed with approved materials, ventilation, and door assemblies. The laboratory layout shown in Figure 1-1 Plan A is optimal because of the greater life safety characteristics afforded by the separate emergency (separate fire-zone) egress. In other layouts, a second egress is permitted through adjacent laboratories, but this arrangement is not recommended.

The major disadvantage of placing vertical service chases at the rear of a laboratory module is that it limits window space within the laboratories. But ventilation plans, energy conservation plans, and efficient operation of laboratory hoods make open windows highly undesirable.

When chases are placed adjacent to a traffic corridor, in contrast to a service corridor, Figure 1-6, maintenance and repairs can be done from the corridor rather than within the laboratory and the second exit from the laboratory can be located away from the exit to the corridor to establish two separate paths of egress. The width of the chase itself forms an alcove against which the laboratory door can swing outward and not project too far into the corridor. Expansive window area on the exterior wall is possible because it will be free of most mechanical services.

The primary disadvantage of locating the chase at an interior corridor wall is that the fume hood, which should be away from the primary egress, will be relatively far from the chase. Therefore, a run of horizontal duct must be used to connect the chemical fume hood, or other special exhaust application, to the riser in the chase. The second exit, although it can be well positioned in the laboratory, may open into another laboratory or office rather than into a corridor in a separate fire zone.

When vertical chases are at an exterior wall, chemical fume hoods can be located adjacent to chases at the rear of the laboratory as shown in Figure 1-5. This arrangement provides a threefold advantage: first, there is a very short horizontal duct connection to the exhaust riser in the chase; second, the fume hood is away from the primary exit; and third, desks can be arranged at the corridor wall away from potential hazards. The need to conduct maintenance to risers and ducts inside chases located within laboratories is a disadvantage. Another disadvantage is that the second exit from laboratories is likely to be near to and open into the same corridor as the primary exit. When chases occupy most of the area of the exterior wall, window area will be restricted.

Locating vertical chases between modules, Figure 1-7, is primarily used for buildings converted to laboratory use. If the structural grid of the building does not conform to the recommended laboratory module dimensions, chases between modules can be sized to make up the difference. When locations for penetrations from floor to floor are limited due to the existing structural system, chases between modules may be the only practical option. A very short horizontal duct connection to the exhaust air riser in the chase will be needed, but the fume hood can be placed away from the primary exit. Window area on the exterior wall will not be restricted. Laboratories with chases between modules are not as flexible as other layouts. They include the disadvantages that pipes and ducts must be serviced within the laboratory and that single-module laboratories will have utilities available along one wall, only.

A central service chase Figure 1-8, has the advantage of a low proportion of mechanical chase area to net laboratory area. Chemical fume hoods can be placed adjacent to the chases at the rear of the laboratories. All pipes and ducts for a cluster of up to four laboratories are accessible from any of the four. Greater variation in module length is possible than for any other layout. The low proportion of mechanical chase area to gross area is an advantage. Because each chase should have space for four dedicated exhaust risers at each floor, there is a limit to the number of floors that a single chase can accommodate unless the chase increases in cross section on upper floors.

1.2.2.5 Structural System

The structural engineer is guided in the design of the structural system of a laboratory building by the organization and dimensions of the laboratory modules on a typical floor. The module lines show where dead loads from walls and benches will occur. Vertical chases for distribution of mechanical services show where major penetrations in the slabs will occur. From data collected on equipment that has been identified through the building program process, the spaces that pose structural or construction problems can be determined. For instance, special design and construction will be required in areas that contain equipment that is unusually heavy or vibration prone. Further, areas that must be isolated from building vibrations will have to be structurally different than the typical floor, as will areas through which it will be necessary to move extremely heavy equipment. Otherwise, structural failure may occur in areas not designed for such loads, even though the load is of short duration.

It is desirable to have as much space as possible above the finished laboratory building ceilings (or equivalent free space above room height, if there will be no finished ceilings) to be used as runs for the large number of ducts, pipes, and electrical services that are characteristic of modern laboratory facilities. Deep solid beams reduce the remaining ceiling clearance beneath.

1.2.2.6 Site Regulations

The site for a new laboratory facility is within the boundaries of municipal, regional, and national jurisdictions. Each government unit has regulations governing land use and construction methods within its boundaries that will affect the proposed facility. Local zoning ordinances, for example, often contain criteria for the following planning concerns: fire district regulations, building use classification, building height restrictions, allowable floor area ratio, clearance and easements around site boundaries, number of parking spaces required on the site, and guidelines on the use of local utilities, such as sewer and water. Some state regulatory agencies require permits for contaminated exhaust air from laboratory fume hoods or local exhaust ventilation systems. Most local zoning and building codes are less restrictive on allowable building height and total floor area when the building is equipped with automatic sprinklers. This manual recommends that laboratory buildings be so equipped throughout.

Approvals by local governing boards and regulatory agencies are generally required before construction can begin. Therefore, a thorough search of all applicable codes, regulations, and ordinances is prudent. Preliminary plan review with agencies that issue critical permits are advised. These discussions allow the owner and designers to anticipate problems in their interpretation of the codes.

1.2.2.7 Building Enclosure

The final objective is to determine alternative schematic designs of building enclosures that will provide the total building area required and comply with the building program and zoning requirements. The term "building enclosure" means the volumetric and geometric characteristics of the structure. This includes perimeter shape, number of floors, building height, site coverage, and orientation to sun and prevailing winds. Building enclosure, total building area, and the structural system of buildings that are under consideration for renovation are known and relatively fixed. Therefore, the assignment of functions to various existing spaces and the proposed modifications to those spaces are done in a manner that will fulfill the building program goals. Because there are fewer available design variables in renovations than in new construction, there is generally less design flexibility.

1.3 GUIDING PRINCIPLES FOR BUILDING HEATING, VENTILATING, AND AIR CONDITIONING SYSTEMS

Ventilation of laboratory buildings is needed to provide an environment that is safe and comfortable. This is accomplished by providing measured amounts of supply and exhaust air plus provisions for temperature and humidity control. Good laboratory exhaust ventilation contains or captures toxic contaminants at the source and transports them out of the building by means of ductwork and a fan. This must be done in a manner that will not contaminate other areas of the building by recirculation from discharge points to clean air inlets, or by creating sufficient negative pressure inside the building to subject hoods to down drafts. The same supply ventilation system may provide makeup air for exhaust air systems plus a comfortable and safe work environment, or there may be a separate supply system for each function. Appendix I contains background information on HVAC system design criteria.

1.3.1 Laboratory and Building Pressure Relationships

Laboratory safety requires a careful balance between exhaust and supply air volumes, as well as concern for the location, design, and quantity of exhaust and supply air. Even within the same type of laboratory, requirements may vary depending on the hazard rating of the materials being used, the quantity of hazardous materials that will be handled, and the nature of the laboratory operations. Communication with laboratory users at an early stage in planning will help to identify and locate the site of potential hazards, making it possible to provide adequate facilities to meet additional needs.

Ventilation systems for laboratories can be divided into three main categories, based on function:

1. Comfort ventilation is provided to the laboratory by a combination of supply and return air flows through ceiling and wall grilles and diffusers. The main purpose is to provide a work environment within a specific temperature, air exchange, and humidity range. Part or all of the comfort ventilation may be provided by special systems installed for health and safety purposes.

2. Supply air systems are required to make up air removed by the health and safety exhaust ventilation systems. Comfort ventilation air may be supplied by one system and makeup air for health and safety exhaust systems by another, or the two supply systems may be combined.

3. Health and safety exhaust ventilation systems remove contaminants from the work environment through specially designed hoods and duct openings.

Components of a ventilation system include fans, ducts, air cleaners, inlet and outlet grilles, sensors, and controllers. Automatic fire dampers are usually required when air ducts pass through fire barriers and are advisable when work with large quantities of flammables is contemplated.

1.3.2 Dedicated and Branched Air Systems

Building supply air may be provided through a central system that serves all areas, for example, offices, storage areas, and public areas, as well as all laboratories. Alternatively, there may be separate systems dedicated exclusively to laboratory usage. Although combined systems are acceptable for supply air, hoods and all other health and safety exhaust facilities should be vented through individual dedicated exhaust ducts and fans. Comfort air exhaust systems may be, and commonly are, combined. Choice of system depends upon budget, space availability, and other considerations such as building size and whether the structure is for mixed use occupancy or solely devoted to laboratory usage.

1.3.3 Constant Volume and Modulating Air Systems

It is advisable, in all cases, to maintain constant pressure relationships between laboratory rooms (greatest negative pressure), anterooms, corridors, and offices (least negative pressure) to avoid intrusion of laboratory air into other areas of the building. For laboratories in which hazardous chemicals and biological agents are used, pressure gradients that decrease incrementally from areas of high hazard to areas open to public access are an essential part of the building's health and safety protective system. Even in laboratories where hazardous materials are not employed, animal holding rooms, animal laboratories, autopsy rooms, media preparation rooms, and similar facilities are likely to generate unpleasant odors. Graded air pressure relationships are usually relied upon to prevent release of foul-smelling air from these rooms. The reverse pressure relationship is required for germ-free and dust-free facilities, such as operating room and white room (clean room) laboratories.

As a rule, in rooms and spaces requiring only comfort ventilation, the pressure relationships relative to the laboratory spaces are intended to be maintained constantly and this calls for invariant airflows. Health and safety ventilation systems may also be designed for continuous, invariant service. Such an arrangement is advantageous when the number of exhaust-ventilated health and safety facilities is small, and the health and safety air supply systems have been combined with the comfort system. However, when health and safety exhaust ventilation represents a major fraction of the total air circulation requirements for the entire building, energy conservation measures call for an ability to shut down these services when not needed. To be effective, there must be two separate supply and exhaust air systems: (1) a comfort ventilation system that provides constant and invariant design temperature, humidity, air exchange, and room pressure conditions, and (2) one or more tempered makeup modulating air supply ventilation systems that are individually coupled to specific exhaust air facilities so both may be turned on and off simultaneously to avoid disturbing the room pressure relationships established by the comfort ventilation system. It is sometimes advantageous to provide a constant volume system to some parts of a laboratory building and modulating systems to others. The nature of the installed facilities and the intensity of laboratory usage will determine the advisability of hybrid ventilation systems.

1.3.4 Supply Ventilation for Building HVAC Systems

1.3.4.1 Supply Air Volume

All air exhausted from laboratories must be replaced with supply or infiltration air. An equivalent volume of replacement or makeup air is essential to provide the proper number of air changes needed to facilitate safe working conditions and to maintain design pressure relationships between rooms and other spaces for health and safety protection. When laboratory health and safety exhaust ventilation requirements are not dominant, total building air conditioning needs for maintaining heating, cooling, and ventilation loads may dictate the supply air volume to each room. Total supply air volume cannot be calculated until the amount of air to be exhausted has been determined.

1.3.4.2 Supply Air Velocity, Temperature, and Discharge Location

The location and construction of room air outlets and the temperature of the air supplied are critical. High-velocity air outlets create excessive turbulence that can disrupt exhaust system performance at a hood face. In addition, comfort considerations make it necessary to reduce drafts. Therefore, the supply air grilles should be designed and located so that the air velocity at the occupant's level does not exceed 50 feet per minute (fpm). There is no single preferred method for the delivery of makeup air; each building or laboratory must be analyzed separately. When supply air is used only for replacing health and safety exhaust ventilation air, as distinguished from temperature control, it is desirable that the air be discharged near the exhaust facility when the space is air conditioned. However, when the room is not air conditioned, there is no advantage associated with close proximity of makeup air to discharge points.

1.3.4.3 Air Intakes

Outside air intakes must be located so as to avoid bringing contaminated air into the building air supply systems. Examination of likely contaminant sources, such as air exhaust points and stacks, should be conducted before outside air intakes are selected. A minimum distance of 30 ft from air discharge openings to air intakes is recommended to reduce fume reentry problems but it is good practice to design for the maximum feasible separation. Outside air intakes located at ground level are subject to contamination from automobile and truck fumes, whereas air intakes at roof level are subject to contamination from laboratory exhaust stacks or high stacks serving offsite facilities in the vicinity. When buildings contain more than 10 stories, it is advisable to locate the air intakes at the midpoint of the building. Difficult sites may require wind tunnel tests to investigate the fume reentry problem under simulated conditions.

1.3.5 Air Discharges

For roof-mounted laboratory hood exhaust installations, the stack on the positive side of the fan should extend at least 10 ft above the roof parapet and other prominent roof structures in order to discharge the exhaust fumes above the layer of air that clings to the roof surface and prevents contaminants from displacing upward. This arrangement will help to avoid reentry through nearby air intake points. To further assist the exhaust air to escape the roof boundary layer, the exhaust velocity should be at least 2500 fpm and there should be no weather cap or other obstructions to prevent the exhaust discharge from rising straight upward (see Figure 2-5).

1.3.5.1 Exhaust Fans

All exhaust fans should be installed on the building roof to maintain the exhaust ducts inside the building under negative pressure as a health protection measure. This arrangement makes it certain that should duct leakage occur, it will be inward. In cases where exhaust air ductwork has to be at positive pressure relative to the building interior, special care must be exercised to ensure by pressure testing that the ductwork is airtight. Many types of exhaust fans are manufactured but only a few meet all the requirements of a good exhaust ventilation system. Double-belted centrifugal utility-type exhaust fans are preferred because they are very reliable, have desirable pressure–volume characteristics, are widely available, and are easily adaptable to roof mounting and attachment of a stack of suitable cross section and height for proper discharge of exhaust fumes. The material of construction for the fan, including protective coatings, should be selected to withstand corrosive and erosive conditions characteristic of the exhaust gases that will be handled. Considerations of life expectancy and maintenance availability will influence the final selection. It is important to specify fans manufactured and rated in accordance with standards established by the Air Moving and Conditioning Association (AMCA).*

Backwardly inclined fans with self-limiting horsepower characteristics have been widely used for general building ventilation purposes. Fan housings are usually constructed of steel and bonderized. When used for exhaust air contaminated with low concentrations of corrosive elements, the interior of the fan and connecting ducts are often coated with a baked primer–finishes especially formulated to meet corrosion resistance standards. For severe corrosive service, especially when high humidity is also present, rigid PVC (polyvinylchloride) or FRP (fiberglass-reinforced polyester) construction is essential. FRP is preferred because of its superior resistance to breakage and vibration cracking. It is necessary to add fire-retardant chemicals to the polyester resin. When the system is located

* 30 West University Drive, Arlington Heights, IL 60004.

on the roof and discharges straight upward without a rain cap, a drain connection should be placed at the bottom of the fan housing. (See Section 2.3.5 for additional design information.)

1.3.6 Supply Air Cleaning

All building supply air, including all portions of recirculated comfort air, should be cleaned according to the requirements of the space. The correct degree of filtration is important because excessive filtration results in a greater pressure drop through the system, thereby increasing operating costs, whereas insufficient filtration results in contamination of critical work areas or excessive maintenance costs from rapid soiling.

Many filter media are available, each providing a specified degree of air cleanliness. The choice depends on the need. American Society of Heating, Refrigeration, and Air Conditioning Engineers (ASHRAE) standards (ASHRAE, 1979) classify filters as throwaway or renewable. Throwaway filters are used once and discarded. They are rated as low efficiency (35% dust removal), medium efficiency (85% dust removal), or high efficiency (95% dust removal). Appendix IV gives the performance characteristics of a number of throwaway dry media filters used for air cleaning. Renewable filters are seldom used for building air cleaning. Electrostatic precipitators are also used for cleaning building supply air. They are designed for 85 or 95% atmospheric dust collection efficiency. Electrostatic precipitators are always made as cleanable units, the interval between cleaning being more or less than three months, depending on the dirtiness of the outside air. Cleaning involves washing the dust-collecting plates with detergent and water. Units may be purchased for manual or mechanical cleaning. Electrostatic precipitators generate small amounts of ozone during normal operation and their use is counterindicated where this gaseous compound would be considered an interference with the work to be undertaken in the new laboratories. Should this be the case, it would also be necessary to remove the same compound from the air introduced into the building, because ozone regularly occurs in outdoor air in most parts of the United States. Ozone, sulfur dioxide, and most hydrocarbons that are normally present in urban air can be removed from ventilation air by passing it through gas-adsorption activated carbon beds after particle filtration or after treatment by electrostatic precipitators.

Cleaning of used comfort ventilation air prior to discharge to the atmosphere is seldom, if ever, attempted, nor is it necessary. This is not always true for health and safety system exhaust air: filtration, liquid scrubbing, and gas adsorption may be needed to prevent the emission of toxic, infectious, and malodorous gases and aerosols to the atmosphere. Exhaust air cleaning will be discussed later (Section 2.3.5.1).

1.3.7 Supply and Exhaust Ducts

All ductwork should be fabricated and installed in accordance with Sheet Metal and Air Conditioning Contractors' National Association Inc. standards (SMACNA, 1985). Ducts should be straight, be smooth inside, and have neatly finished joints. All ducts must be securely anchored to the building structure. The usual material for supply and exhaust ducts in comfort ventilation systems is galvanized steel. More corrosion-resistant materials, such as stainless steel, fiberglass–reinforced polyester, and PVC, are frequently used for health and safety system exhaust ducts.

It is essential that all ducts be constructed and installed in a leak-tight manner if the system is to function as the designer intended it to. This is especially important for exhaust ventilation ducts, which usually operate under far higher negative pressures than comfort ventilation systems; hence the leakage through even small gaps in longitudinal seams and joints leads to a significant drop in system efficiency. The seams and joints of stainless steel ducts are usually welded airtight. Plastic pipes are constructed without longitudinal seams and the joints are sealed with plastic cements of appropriate composition. It is easier to construct airtight systems with round than with rectangular ducts. Rectangular ducts should be avoided at any cost in the health and safety exhaust air systems because they cannot be made airtight by any practical method. For noncorrosive material-containing systems, seams and joints may be sealed with long-lasting ventilation duct tape. Whatever method is selected, it is essential that ducts be made leak-tight if they are to give satisfactory service.

The current poor state of ventilation system engineering and installation is exemplified by the inclusion of redundant "trimming dampers" in all duct runs and the custom of engaging ventilation system "balancing specialists" to place the systems into full compliance with design airflow values after the building is completed, instead of carefully designing a balanced system (without dampers) and making certain that it is constructed and installed in strict accordance with the design specifications. Although it is customary to balance ventilation systems after the building is completed, it is often the last construction step. It would be much better to test and balance the system as early as possible or, in any case, before the duct systems are concealed above false ceilings and behind cabinets, hoods, and similar large pieces of laboratory furniture.

Ducts are excellent conductors of sound. Special care must be expended to secure them so as to avoid vibration. In addition, they should be isolated from fans and other noise-generating equipment with the use of vibration-reducing connections and the installation of acoustical traps in the ducts between noise and vibration sources and the point of discharge to occupied areas.

1.3.8 HVAC Control Systems

Controllers are essential to ensure that all HVAC systems in a laboratory building will operate in a safe and economical manner. Control systems are needed for temperature, humidity, air exchange, and pressure regulation.

1.3.8.1 Temperature Control

The temperature in most laboratory buildings does not require close regulation, that is, no better than ± 3°F. Worker efficiency and productivity may be affected adversely when ambient temperature control permits temperatures to exceed 85°F in summer or to fall below 60°F in winter.

The comfort range is determined by combining dry bulb temperature, relative humidity (RH), and air velocity to derive a value called effective temperature (ASHRAE, 1977). An effective temperature of 77°F is considered to be a very desirable condition. This is achieved in winter by 68°F with 35% RH (± 5%), and in summer by 78°F with 50% RH (± 5%).

1.3.8.2 Humidity Control

Although close humidity control is not required in most buildings, some degree of humidity control should be included to provide comfort and avoid extremes.

1.3.9 Air Exchange Rates

Recommended air exchange rates for public areas and for commercial and industrial workplaces are contained in the ANSI-ASHRAE draft ventilation standard 86-62 (ANSI-ASHRAE, 1986). In most organized communities, the applicable building code will prescribe minimum ventilation rates for a variety of building usages. Where they exist, they will take precedence. They are the minimum air exhaust rates that must be maintained in each occupied room, even when the health and safety exhaust ventilation systems are turned off.

In crowded areas where smoking is permitted, as in waiting rooms, eating places, and group offices, the minimum air exchange rates required by the building codes will be inadequate to please a high percentage of nonsmokers and outside air rates of 30 cfm (cubic feet per minute) per active smoker will be needed to keep tobacco smoke concentrations close to background levels (Yaglou, 1955).

1.4 LOSS PREVENTION, INDUSTRIAL HYGIENE, AND PERSONAL SAFETY

1.4.1 Emergency Electrical Considerations

The primary electric feed to laboratory buildings should be as reliable as possible. For example, separate and distinct feeds connected to a common bus and then to two separate transformers with network protectors should be installed, and each transformer should be large enough to carry the building load so that loss of any one line will not interrupt building power. When such practices are not possible, other fail-safe electrical connection designs should be used. Even with this type of reliable service, it is often necessary to provide emergency electrical power because any one of the primary electrical service components (transformers, main feeders, etc.) may fail and then emergency power will be required. Each building should have its own emergency power source that is adequate for all egress lighting and other life-safety requirements, as outlined in NFPA 70 and 101, (NFPA 70, 1985; NFPA 101, 1985). Several critical systems in laboratories may have to be connected to emergency electrical power for continuity of operation as well as for safety concerns. Items that should be connected to emergency electrical power are the following:

1. Fire alarm systems.
2. Emergency communication systems.
3. One elevator for buildings over 70 ft in height.
4. Egress lighting.
5. Emergency lighting in rooms.
6. Egress signs.
7. Fire pump, when it is electrically driven and not backed up by another driver.
8. Exhaust fans connected to critical health and safety exhaust ventilation systems.
9. Makeup air systems connected to critical exhaust systems.
10. Heating system controls to prevent building from freezing during temperature extremes.
11. Emergency smoke evacuation systems.

12. All other systems whose continuing function·is necessary for safe operation of the building or facilities during an emergency period.

In general, diesel-driven generators are preferred because they are readily available, easily maintainable, and easy to start (in less than the 10 s mandated by NFPA 70). Gas turbines are available in smaller sizes and may be satisfactory. However, starting is sometimes difficult for the large sizes.

The emergency generator should be connected to the selected load with a series of transfer switches. The transfer switches should automatically turn over to emergency power when normal power fails. Annunciation through a local or remote panel should be provided to let operators of the building know which transfer switch has changed modes. The generator control board should have an ammeter installed so operators can see the load on the generator and manually select other loads when necessary.

If emergency electrical power distribution is run throughout the facility, it must be run on a distribution system that is separate and distinct from the primary electrical distribution system. This is to prevent concurrent cable failure of both primary and emergency power·in case of a fire or other emergency condition.

1.4.2 Construction Methods and Materials

According to the Building Officials and Code Administrator's International Basic Building Code (BOCA, 1985), laboratory buildings engaged in education, research, clinical medicine, and other forms of experimentation are included in the category of Class A building usage. Therefore, all pertinent sections of the BOCA code should be followed in the design and construction of all laboratory facilities, with special emphasis on fire safety for unusual as well as all ordinary hazardous conditions. Specifically, the provisions of Article IX (Fire-Resistive Construction Requirements) that govern the design and use of materials and methods of construction necessary to provide fire resistance and flame resistance must be followed. Flame resistance is defined in the BOCA code as "the property of materials, or combinations of component materials, which restricts the spread of flame as determined by the flame-resistance tests specified in the Basic Code." Some of the specific subjects covered in Article IX of the BOCA code are enclosure walls, fire walls and fire wall openings, vertical shafts and hoist ways, beams and girders, columns, trusses, fire doors, fire windows and shutters, wired glass, fire-resistance requirements for plaster, interior finish and trim, and roof structures. The purpose of these requirements of the BOCA code is to provide a building that will allow its inhabitants to move freely from the building in case of unusual fire and/or smoke conditions. Additional chapters of the code address other aspects of life safety.

The general philosophy of all interior building design with respect to the combustible properties of construction materials should embrace the idea of eliminating those materials responsible for rapid flame spread and heavy smoke generation. Materials frequently used in research buildings provide more than ordinary cause for concern, especially because the sources of fire initiation in laboratories are many times greater than for most other building uses.

1.4.3 Safety Control Systems for Laboratory Experiments

Provisions should be made for automatic or remote shutdown of well-defined portions of a building's services that provide energy to experimental operations having the capability to threaten parts of the building or personnel within the building, or to produce undesired effects should the experiment get out of hand while attended or unattended. This type of control should be used for the most sensitive types of operations where uncontrolled failure could result in a major loss of building or equipment, the release of highly toxic substances into the environment, or personal injury. A study committee composed of designers, users, and representatives of the health and safety professions should be set up to determine areas of need that will benefit by application of this requirement.

1.4.4 Fire Detection, Alarm, and Suppression Systems

Costs of installing fire detection and suppression systems are very high after the construction of any type of laboratory building. Therefore, consideration should be given to this need during the design stage of new and renovated laboratories. Sprinklers are considered the best fire control.

1.4.4.1 Fire and Smoke Detection

All laboratories should have a heat-sensitive detection system in addition to a standard sprinkler system with a slower detection quality. Because most ionization detection systems sense products of combustion other than heat, their use in laboratories generating volatile chemicals, using open flames, and equipped with devices that require fairly high velocity air movement is not recommended. Ionization detectors are prone to sense fires that are not a threat and to fail to see threatening fires sooner than more stable thermal detectors. Several LED-operated (light emitting diode) visual smoke detectors are available which can detect smoke at levels below 4% obscuration. If very early warning of smoke is desired, they would be appropriate units to use. For reliability reasons, incandescent lamp detectors should be avoided. The rate of temperature-rise fire detectors are probably the best and most reliable of the available devices but when compared with twisted pair thermal detectors, they do not cover as much area per unit of cost. However, twisted pair thermal detectors are prone to damage through improper use and cost more to repair and reset after fire or damage. (We have seen investigators hang materials from the wires as one would from a clothes line, causing them to short and sound an alarm.) Fire detectors must meet the Underwriters Laboratory (UL) standard and be installed and spaced in accordance with NFPA 72E (NFPA 72E, 1985).

1.4.4.2 Fire Suppression

1.4.4.2.1 Fixed Automatic Systems

All laboratory buildings should be designed with a complete water sprinkler system in accordance with NFPA 13 (NFPA 13, 1985). Wherever unusual hazards exist, special design of the system will be necessary. When water is contraindicated for fire suppression because of the presence of large amounts of materials such as elemental sodium that react violently with water, other complete fire suppression systems must be used. They include Halon 1301 for electronic, high-voltage, or computer laboratories, and total flooding with dry chemical for chemical storage areas. A competent fire protection or safety engineer should participate in these decisions. All automatic fire suppression systems should be connected to the building central alarm system.

The vertical standpipes used for the water sprinkler system should also serve fire hose cabinets on each floor of the laboratory building. Hose size should be $1\frac{1}{2}$ in. diameter and vertical risers should be so spaced that the maximum length of hose to reach a fire will not exceed 50 ft. Longer hose runs may lead to loss of fire control because a hose length exceeding 50 ft is difficult to turn on and use by persons lacking hands-on fire training.

1.4.4.2.2 Hand-Portable Extinguishers

Hand-portable fire extinguishers should be located in halls and main exitways, as well as within individual laboratory units. A clean fire extinguishing agent, such as carbon dioxide (CO_2), is usually preferred for use in laboratories. Within the laboratory, 5-lb units are the minimum recommended size.

Fire extinguishers installed in halls and exits should be sized and installed in accordance with NFPA 10 (NFPA 10, 1985). The larger units will generally be multipurpose dry chemical extinguishers that will be more suitable for fighting a large laboratory fire. Small fires in the laboratory can be handled adequately with a CO_2 unit without having to cope with the mess and required cleanup of dry chemicals.

1.4.5 Alarm Systems

1.4.5.1 Fire Alarms

A Class A supervised fire alarm or signaling system should be installed throughout the laboratory building in accordance with NFPA 72A (NFPA 72A, 1985). It should have all manual pull stations, sprinkler alarms, and detection systems (heat, fire, or smoke) connected to it.

1.4.5.2 Laboratory Experiment Alarms

Provisions should be made for a three-tier alarm system in all laboratories where experiments or operations need to be monitored for failure. The system should be designed to provide a communications link between the laboratory and a central station in the building that is monitored at all times, or, at the very least, when there are unattended operations in laboratory units. In general, a three-tier alarm system consists of the following parts:

1. A local alarm for room occupants that is audible and visual.
2. An audible alarm outside the laboratory door to pinpoint the location of the problem.
3. Remote annunciation to a constantly attended location.

The use of remote annunciation is critical in a large facility because it may be the only means of alerting service personnel to the problem. Remote annunciation is most critical during weekends and normally unoccupied periods. It is strongly recommended that alarms to all electromechanical equipment connected to laboratory safety systems be annunciated to a central location.

The use of microcomputer technology may be advantageous for the designer to consider when more than one kind of alarm condition must be monitored, for example, fire alarms and HVAC alarms.

1.4.5.3 Other Service Alarms

Alarms may be needed to indicate failure of exhaust fans and makeup air systems, as well as for fire, loss of pressure, loss of temperature, presence of toxic gases, and other conditions that often require monitoring. In addition, whenever building services that are not normally monitored could cause loss through flood, fire, explosion, or release of hazardous materials in the event of their failure, a separate monitoring system with three-tier alarms should be installed. Alternatively, equivalent facilities could be added to the laboratory experiment monitoring and alarm system.

1.4.6 Hazardous Waste

An area must be provided to collect and store chemical wastes in preparation for disposal. General waste collections, such as those from janitors' operations, should consist solely of paper, glass, and other nonhazardous refuse, but should not include waste chemicals. General waste should be collected and stored in an area of the building not associated with the chemical waste storage area. The chemical waste storage area should be capable of being within the main laboratory building or in an external facility.

Chemical waste storage rooms should be located at or above grade and on an outside wall containing explosion venting glass windows in a surface-to-volume ratio of at least 1 ft² of window area for each 40 ft³ of room volume (NFPA 68, 1985). The room should be protected with a fire-supression system that includes total flooding or dry chemical actuated by flame, pressure, or heat (but not ionization) detectors. The room should be sized to handle the anticipated buildup of materials in accordance with the facility's disposal volume and frequency of collection. Excellent exhaust ventilation should be provided with a minimum of 10 air changes per hour to purge fumes from leaking or spilled packages.

The room should be divided to make possible separation of incompatible chemicals. Only explosion-proof electrical outlets meeting Division 2 NFPA 70 concepts (NFPA 70, 1985) should be used for all electrical equipment. Grounding and bonding methods must also be provided for all waste solvent transfer operations involving 5-, 30-, and 55-gal drums. Solvent transfer areas where solvents are exposed to the atmosphere must also meet requirements of Division 2 of NFPA 70.

At least one 6A 60BC hand portable fire extinguisher should be located inside the room at an easily accessible location. A fire blanket, emergency eyewash station, and a deluge safety shower should be located within or just outside the room. Local exhaust ventilation in the form of a chemical fume hood or a similar facility may be needed inside the room for handling toxic materials or noxious animal wastes. Carcasses may require the installation of a freezer which is UL tested as explosion proof for the highest class of flammable hazard anticipated—probably Class I Groups C and D.

1.4.7 Chemical Storage

In addition to providing for the storage of a few day's supply in each laboratory unit, a central chemical storage room should be provided. This room should be sized to hold enough materials to assure continued operations without interruption. The purchasing department can assist in determining this quantity. Compressed gas cylinders should not be stored in this room unless there is a subroom with high rates of ventilation. Good floor drainage should also be provided for compressed gas storage areas.

Shelves, cabinets, drawers, hoods, and special storage equipment should be installed to provide separation of incompatible chemicals. Shelves should have a $\frac{3}{4}$- to $\frac{1}{2}$-in. lip on the edges to prevent containers from falling off due to vibrations. Four types of storage should be provided:

1. Shelves for nonhazardous chemicals.
2. Cabinets for controlled substances and very toxic materials.
3. Cabinets with fire ratings for storage of flammable liquids (see "Steel Storage Cabinets for Flammable Liquids," *Factory Mutual Approval Guide 1982* (FM, 1982) and OSHA 1910.106 (OSHA, 1978) for steel and wood cabinet requirements).
4. Fume hoods or other exhaust-ventilated facilities for unsealed toxic chemicals.

If operations in the chemical stockroom will involve the transfer of flammable liquids from container to container, the requirements of the National Fire Protection Association (NFPA 30, 1985) must be met. These include provisions for bonding and grounding of containers and explosion-proof electrical equipment.

The chemical stockroom should be protected with a standard water sprinkler system except where large quantities of flammable liquids or water-reactive chemicals dominate the stockroom. In those cases, an appropriate system, such as total flooding dry chemical, should be used. A 6A 60BC fire extinguisher, a fire blanket, and an emergency eyewash facility should be located within the stockroom and a deluge safety shower should be nearby in a hall or corridor. For additional reading on chemical storage, see *Safe Storage of Laboratory Chemicals* (Pipitone, 1984) and *Hazardous and Toxic Materials* (Fawcett, 1984).

1.4.8 Compressed Gas Storage and Piping

When the laboratory management elects to pipe gasses from a central compressed gas dispensing facility instead of placing cylinders in each laboratory unit, the location of the central facility and an outline of the design features must be included as an integral part of the building design.

When a central tank farm is used, it should be located in a room with an outside wall or be housed in a room attached to the outside of the building. The room should be adequate in size to store in segregated locations (1) full cylinders and (2) empty cylinders awaiting removal for refilling, as well as the manifolds necessary for piping the gasses. Ventilation in this room should be adequate to vent heat from the sun load on the roof and walls and to remove gases leaking from a failed regulator or valve. Air changes should be a minimum of 10 per hour.

Rigid and secure supports for gas tanks should be provided and they should be designed to provide storage flexibility. Compressed gas cylinders for a high-pressure laboratory would be likely to be located within that laboratory, or close to it, to avoid any loss of high discharge pressure in the piping system that occurs for the general laboratory gases when piped from a central location.

The gas piping system should be of stainless steel with low pressure reducers at the gas tank farm and orifice restrictions wherever pipe diameter exceeds $\frac{1}{4}$ in. to limit accidental flow into any area. The piping system should be external to the building when this is feasible. Internal piping and exterior piping, alike, should be exposed to view wherever possible. Excess flow check valves may also be installed to control runaway flow conditions of toxic or flammable gases (see Appendix IX).

1.4.9 Fuel Gas

A fuel gas shutoff must be provided for the entire building.

1.4.10 Hazardous Materials, Equipment, and Procedure Signs

Personnel within a lab building or about to enter a lab building need to have some information regarding the operations, materials, risks, or special situations within. This information is most important to emergency response personnel, such as fire fighters and police or ambulance personnel, so that they can carry out their functions safely, usually in time of stress. Many communities and cities have ordinances requiring certain signs such as those for flammable storage, which must be complied with. One system is described in NFPA 704 (NFPA 704, 1985), which was developed around nonlaboratory users of chemicals. You may find an ordinance requires compliance with 704. If not we recommend adoption of some other system that can protect emergency response personnel in laboratory situations. Such a system is shown in Appendix X.

1.5 MISCELLANEOUS SERVICES

1.5.1 Lighting Power Limits

A lighting power limit is the maximum power that may be available to provide the lighting needs of a building. Separate lighting power limits should be calculated for the interior and exterior of the building. To establish lighting power limits, the following procedure should be used. The intent here is to impose energy conservation discipline to minimize overlighting or overusage of the electrical power.

For Interiors

1. Determine the use categories for the various parts of the building from Table 1-5.
2. Multiply the maximum power limit for each category by the gross floor area included in that category.

TABLE 1-5
Lighting Limit (Connected Load) for Listed Occupancies

Type of Use	Maximum W/ft²
Interior	
Category A: Classrooms, office areas, mechanical areas, museums, conference rooms, drafting rooms, clerical areas, laboratories, kitchens, examining rooms, book stacks, boiler rooms, combined kitchen and dining facilities, libraries, valence and display case lighting	2.00
Category B: Auditoriums, waiting areas, restrooms, dining areas, working corridors, book storage areas, active inventory storage, stairways, locker rooms, filing areas of offices, shipping and receiving areas	1.00
Category C: Corridors, lobbies, elevators, inactive storage areas, and foyers	0.50
Category D: Indoor parking	0.25
Exterior	
Category E: Building perimeter: wall-wash, facade, canopy	5.00 (per linear foot)
Category F: Outdoor parking	0.10
Lighting switching: In all areas exterior to the building lighting fixtures should be capable of being switched automatically for nonoperation when natural light is available.	

3. Add the total number of watts for each area to arrive at the total lighting power limit for the building.
4. In open-concept office spaces in excess of 2000 ft² without defined egress or circulation pattern, 25% of the area can be designated as category B in Table 1-5.
5. In rooms with ceiling height in excess of 20 ft, a power allowance, in watts per square foot, of an additional 2% per foot of height can be used, up to a maximum of twice the limit in Table 1-5.

For Exteriors
6. Facade lighting: multiply the limit given in Table 1-5 by the number of linear feet in the building perimeter.
7. Parking: multiply the value in category F in Table 1-5 by the area to be illuminated.

Allowable Exceptions
8. Task lighting need not be included in the lighting power limit.
9. Process lighting in the laboratories need not be accounted for in the lighting calculation.

1.5.2 Lighting Level Guide

Suggested minimum lighting levels are presented in Table 1-6. These values can be compared with those calculated from Table 1-5. If large differences occur, the level of choice will depend on room occupancy.

1.5.3 Plumbing

1.5.3.1 Sinks

Sinks should be constructed of materials such as stainless steel or epoxy-coated resins which are resistant to possible chemical and other spillage. The drain should have a removable, cleanable strainer to prevent solid materials from getting into the drainage system.

1.5.3.2 Liquid Wastes

In some laboratories, special waste-handling systems will be necessary, including dilution tanks under sinks, a central building dilution tank, a monitoring system to measure the pH of all waste and automatically add acid or caustic to control the pH before it enters the municipal sewer system. An accurate chart log of the pH before and after treatment will

TABLE 1-6
Lighting Level Guide

Operation of Area	Suggested Minimum[a] Lighting Level (Foot-Candles on Work Task)
Administration Areas and Offices	
Cartography, detailed drafting, designing	100
Accounting, bookkeeping, tabulating, business machines	60
Executive offices, general offices, reading and transcribing pencil handwriting, active filing, mail sorting	60
General offices with easier visual tasks, intermittent filing, mechanical reproduction	50
Conference rooms, inactive file areas, general reading of books, periodicals	30
Laboratory areas	
Microanalytical, critical or delicate operations, close work, etc.	70
General analytical, routine analytical, physical testing	50
Engine laboratories, equipment test areas, fume hoods	30
Pilot plant and process areas (Indoor)	
Process equipment	30
Main operating aisles	30
Sampling points	30
Feed and product handling	30
Secondary aisles	20
Control panel	50
Console or desk	90
General building areas	
Stockrooms, reception areas	30
Corridors, hallways, stairways	20
Washrooms, toilets	20

[a] The "suggested minimum" values will permit good work performance in almost all cases, but lighting levels much below these values will cause a reduction in visual performance and accuracy. Consult company industrial hygienist if higher light levels are needed. *Note.* Values shown are for in-service conditions. Initial performance figures should be one-third higher.

assure compliance with regulations of discharge. In other locations, it may be possible to install a central collecting station whereby all such waste is accumulated and then reclaimed. Sometimes, it may be necessary as a matter of policy for occupants in the laboratory to put waste into a reusable container.

1.5.3.3 Water Pressure

Sufficient water pressure should be available for all the building needs. Separate piping loops are necessary for the following: sprinkler system, potable water, laboratory water, eyewash fountains, and deluge showers. Antiscalding temperature regulating devices must be installed in service hot water supply lines. For eyewash and deluge shower specifications, see Appendixes VII and VIII. For standpipes in locations where municipal water supply does not provide sufficient pressure, separate water pressure booster systems are necessary. In locations where municipal water supply is not present, or the quantity or quality is not sufficient, separate water storage systems may be necessary. For example, a laboratory building being built in Saudi Arabia may require a complete water system.

1.5.3.4 Drinking Water Protection

Laboratory buildings require protection of their drinking water systems to prevent contamination from laboratories. This requires separation of the laboratory water system within the building from the water systems used for drinking, kitchens, toilet rooms, emergency showers, and eyewashes. The need to conserve and protect entire municipal drinking water supplies from contamination due to backsiphonage or backpressure has been recognized, for example, by the Massachusetts Department of Environmental Quality Engineering (DEQE) regulation 310 CMR 22.22 (CMR, 1979). The regulation describes the need and the approved method for protecting state, city, town, and local drinking water systems from any possible degradation caused by cross-contamination. A double check valve, reduced pressure backflow preventer with a relief valve and open drain is the only acceptable method approved by regulation 310 CMR 22.22. It offers the best protection of all the backflow preventers available and can be used on toxic and nontoxic systems.

Testing of backflow preventers is required semiannually. An additional reduced pressure backflow preventer installed in a bypass arrangement is required to enable these tests to be done without loss of water service to the building.

The drinking water system contained within the building should be protected in the same way as the municipal supply, by using reduced pressure, backflow preventers and a bypass line to avoid loss of service during semiannual testing. The domestic hot water supply system requires similar treatment to provide the same kind of protection to building occupants.

1.5.4 Support Services

When designing laboratory buildings it is important to consider the health and safety issues related to laboratory support service personnel, for example, maintenance, housekeeping, etc. Adequate space must be provided for housing these people and their equipment. These areas should be provided with the same health and safety features as the other laboratory areas, for example, ventilation, fire protection, egress, and so on. In addition there may be some special considerations. For example, non-slip floor surfaces should be provided in areas where large quantities of water may be on the floor frequently, i.e., glasswashing room.

Adequate ventilation and work space must be provided in mechanical and fan rooms. Be aware that routine maintenance will need to be performed on laboratory ventilation systems and proper access must be provided.

2

Laboratory Considerations

2.1 GUIDING CONCEPTS

All laboratories, regardless of their specific use, have many similar health
and safety requirements that should be considered in the design stages.
This chapter reviews the requirements that are common to all types of
laboratories discussed in this text but does not deal with specific laboratory types. In Part II specific laboratory types will be discussed and the
distinguishing features of a number of commonly encountered laboratory
types will be described and illustrated. In some cases, specialized laboratories may not require one or more of the items discussed in this section;
in other cases, it will be necessary to describe special facilities not discussed in this chapter; and sometimes it will require a combination of
both. Unless otherwise noted, items discussed here will be referred to by
section number when specialized laboratory types are covered in Part II.

2.2 LABORATORY LAYOUT

The laboratory layout is critical for the efficient use of space and the
safety of laboratory personnel. The laboratory design must be consistent
with the building design philosophy described in Chapter 1. This includes
provisions for entry and egress, furniture and equipment locations, and
access for handicapped persons. Consideration of safety issues in the

planning and schematic design phases of a project saves the owner/user the cost of corrective modifications of the plans and materials during construction or the continued liability of built-in safety hazards. Neglect of safety considerations in the design phase can lead to laboratories containing needless hazards to health and safety.

Alternative laboratory designs, with regard to health and safety considerations, are summarized in Figures 2-1 through 2-5.

2.2.1 Personnel Entry and Egress

There are many specific requirements for egress from laboratory spaces. The major reference sources for safety standards and codes are those published by the National Fire Protection Association (NFPA), U.S. Occupational Safety and Health Administration (OSHA), and Building Officials and Code Administrators, International Inc. (BOCA). They set policies on the following items: the number of exits, direction of door swing, and permissible door swings into egress pathways. The essential features are summarized below.

A minimum of two exits from each laboratory is required by NFPA 45 (NFPA 45, 1985) and, under certain conditions, by OSHA regulations (OSHA 1910.37, 1978). The safest arrangement for a laboratory is to have each exit open into a separate fire zone. The exits should be located to provide separate paths to evacuation routes. When a hazard makes one egress corridor impassable, the second provides an alternative safe route out of the building.

Exit doors should swing in the direction of egress because an outswinging door cannot be blocked by persons being pushed against the door by those in a panic behind them. Also, in an emergency, it takes less time to push open an outswinging door than a door which swings inward. Finally, a solvent fire in a tightly constructed laboratory room might raise room pressure sufficiently to make it difficult to open the door.

To make certain that egress corridors will not be blocked by open doors or other obstructions, outswinging doors should be recessed sufficiently that the door does not protrude more than 7 in. into the clear width of the corridor when fully open. This makes it impossible for traffic passing in the corridor to block the opening of the laboratory exit door, nor will persons passing be hit as the exit doors open.

Glass panels of 100 in.^{2} or less are permissible in fire-rated laboratory exit doors. They help to prevent collisions of persons entering and exiting. The glass should be placed low enough that persons of less than standard height or in wheelchairs can be seen from the other side of the door.

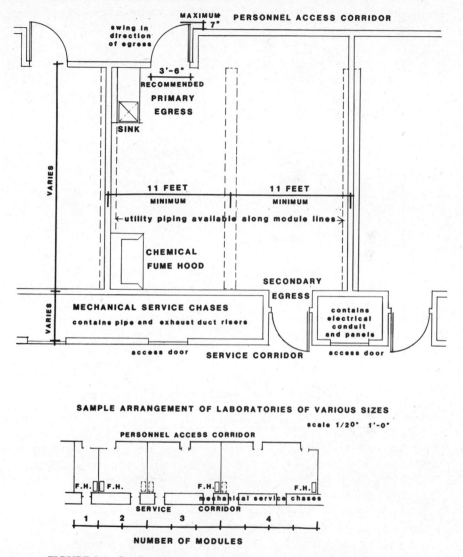

SAMPLE ARRANGEMENT OF LABORATORIES OF VARIOUS SIZES

scale 1/20" 1'-0"

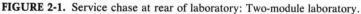

NUMBER OF MODULES

FIGURE 2-1. Service chase at rear of laboratory: Two-module laboratory.

Many codes concerned with removal of architectural barriers in buildings specify lever action handles for workplace doors because levers are easier to activate than knobs in emergencies. Further, the firm downward motion to release the latch does not require the use of hands. Architectural barrier codes also limit the pressure required to open doors containing automatic door closer mechanisms.

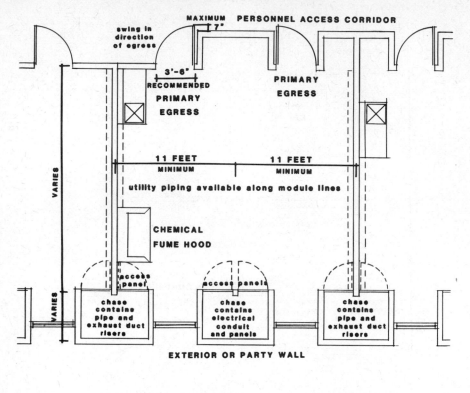

FIGURE 2-2. Service chases at exterior wall: Two-module laboratory.

The minimum dimensions for exit doors are 32×80 in., but standard practice and architectural barrier codes recommend a minimum of 36×80 in. to facilitate the movement in and out of laboratories of persons in wheelchairs, as well as equipment and carts. Openings that do not qualify as exits include windows, trap doors, and knock-out panels between laboratories. Doors onto the balconies of multistory buildings are not considered to be legal exits. Exits must be marked in accordance with local and national standards.

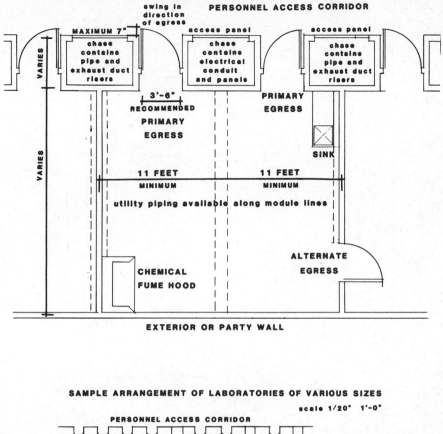

FIGURE 2-3. Service chase at corridor wall: Two-module laboratory.

2.2.2 Laboratory Furniture Locations

Laboratory benches, desks, and other furnishings must be designed and located to facilitate ease of egress and ease of travel within the laboratory. Major aisles between benches, whether located against walls or in the center of the room, should be aligned in the direction of egress.

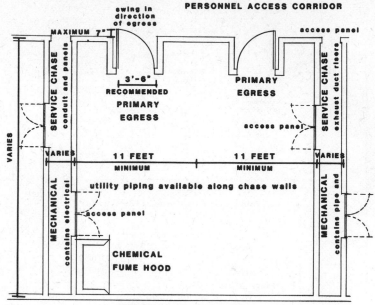

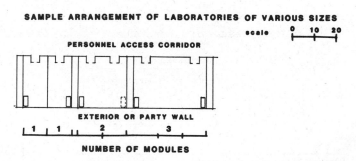

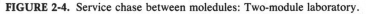

FIGURE 2-4. Service chase between moledules: Two-module laboratory.

2.2.2.1 *Benches*

Center room benches may be of the island type with aisles on all sides making it possible for personnel to move around the bench quickly to obtain emergency equipment or to exit. If there is an approved exit on both sides or at both ends, a peninsula type bench is acceptable. Single-module laboratories will usually be equipped with benches along both long walls and have a central aisle between them.

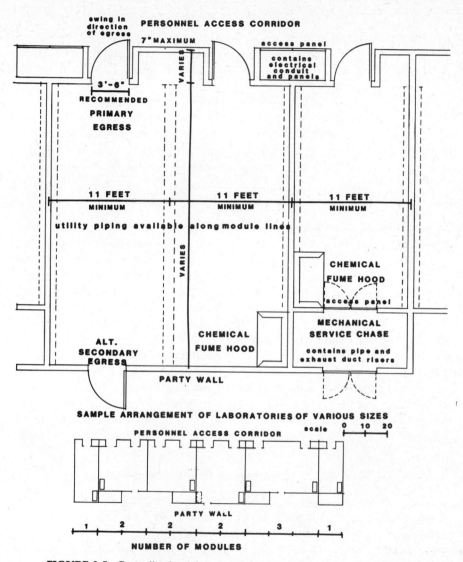

swing in direction of egress

PERSONNEL ACCESS CORRIDOR

7" MAXIMUM

VARIES

access panel

contains electrical conduit and panels

3'-6"
RECOMMENDED
PRIMARY
EGRESS

11 FEET
MINIMUM

11 FEET
MINIMUM

11 FEET
MINIMUM

utility piping available along module lines

VARIES

CHEMICAL
FUME HOOD

access panel

ALT.
SECONDARY
EGRESS

CHEMICAL
FUME HOOD

MECHANICAL
SERVICE CHASE

contains pipe and exhaust duct risers

PARTY WALL

SAMPLE ARRANGEMENT OF LABORATORIES OF VARIOUS SIZES

scale 0 10 20

PERSONNEL ACCESS CORRIDOR

PARTY WALL

1   2   2   2   3   1

NUMBER OF MODULES

FIGURE 2-5. Centralized service chase: One- and two-module laboratories.

2.2.2.2 *Aisles*

Major aisles between benches or equipment should have a minimum clearance of 5 ft to allow for safe passage of persons behind those working at a bench (Nuffield, 1961). (See also Section 1.2.2.3.1.)

2.2.2.3 *Desks*

If desks are needed in the laboratory, they should be placed so that they are away from potentially dangerous laboratory operations and located so that the path from the desk to the means of egress does not require movement past or in a direction of increasing hazard. (See also Section 1.2.2.4.)

2.2.2.4 *Work Surfaces*

Work surfaces and base storage units should be dimensioned to permit safe access to utility outlets and easy reach to the storage units above the work surface. The standard work surface depth is 2 ft not including the utility strip. Deeper surfaces may be needed for large pieces of bench-mounted equipment, but safe access to utilities must be maintained. If the clearance between the horizontal work surface and the underside of a wall-mounted storage unit is less than 2 ft, projection of these units over the work surface should not be greater than 1 ft. With this geometry, it is not likely that intense heat sources or open flames will be placed beneath the shelf and cause a fire. Work surface heights in the laboratory are designed for two body positions: seated work surface height is between 30 and 32 in. and standing work surface height is between 35 and 37 in. Seated work stations are advised for all laboratories where handicapped personnel may work and where microanalytical techniques may be used.

2.2.3 Location of Fume Hoods

Because hazardous operations are usually conducted in the laboratory hood, chemical fume hoods must be located away from the primary laboratory exit for safety. In addition, placement at the far end of a laboratory aisle reduces traffic in front of the hood, and this, in turn, reduces adverse effects on hood performance. (See also Section 1.2.2.4 and Appendix V.)

2.2.4 Location of Equipment

Layout of fixed and movable equipment is determined by the hazards presented by the equipment, requirements for utility connections, and the area the equipment occupies. The most hazardous processes and associated equipment should be placed away from the primary means of egress.

Equipment posing little or no hazard can be placed closer to the egress route. Considerations of area requirements and utility connections are listed in Table 2-1. (See also Section 1.2.2.4.)

2.2.5 Handicapped Access

All parts of the laboratory and its emergency equipment should be accessible to handicapped persons. Ramps, eyewash locations, deluge showers, fire alarms, electrical outlets, switches, and controls should be designed in accordance with applicable achitecturier barriers codes. State codes differ: some states have none, whereas the codes of California, New York, and Massachusetts are among the most detailed.

TABLE 2-1
**Checklist for Selection and Location of Laboratory Furnishings
and Equipment**

1. List all major pieces of equipment in each laboratory according to the following characteristics:
   (a) Size and weight
   (b) Whether bench mounted, free standing, or other mounting method
   (c) Fixed to wall or floor with permanent utility connections
   (d) Moveable with quick disconnects to utilities
   (e) Required utilities and services
      (1) Water: city, purified, recirculated process
      (2) Drain: floor, piped
      (3) Gas: piped, manifolded tanks
      (4) Compressed air
      (5) Vacuum
      (6) Electricity: voltage, amperage, phases, location and number of standard and special receptacles
      (7) Local exhaust services: type, volume, filters, other treatment
      (8) Air supply for local exhaust services: tempered, filtered, humidified
      (9) Steam: pressure, volume, flow rate
   (f) Maintenance clearances
   (g) Operator's position and clearances
   (h) Equipment attachments
   (i) Heavy rotary components
2. Determine the location of and area required for supplies associated with each piece of equipment:
   (a) Glassware
   (b) Instruments, manipulators
   (c) Chemicals
   (d) Disposables: paper goods, plastic ware

2.3 LABORATORY HEATING, VENTILATING, AND AIR CONDITIONING (HVAC)

Laboratory ventilation is closely related to correct overall laboratory building function in the sense that laboratory HVAC services will be enhanced or constrained by the design choices made for the building as a whole. Prudent design will take into account the unusually rapid changes that are presently occurring in most research areas. Often they require a greater number of instruments and other equipment per program or per project to support new methods of experimentation. By the very nature of research, old restraints on pressure, chemical composition, toxicity, etc. are constantly being cast off and new demands placed on the provisions of essential facilities. Therefore, reserves should be designed into all laboratory HVAC systems to retard obsolescence. A reserve of 10–25% of new requirements is recommended in all cases although for industrial systems 35–50% excess is more usual.

2.3.1 Temperature Control

Heating and air conditioning systems must provide the uniform temperature that is needed for correct operation of many analytic devices. Although close control of humidity will not be necessary in most instances, when close humidity control is required, simultaneous heating and cooling may be needed. Many kinds of HVAC systems have been used for laboratories with success: local cooling units, central systems, or a combination of both. It is important that all elements of the temperature control systems be interlocked so that a uniform temperature can be maintained throughout the year. Often it is advantageous to install a separate air cooling and heating system for a laboratory rather than connecting it to the building system because laboratories have HVAC requirements that are different from those of a normal office or hospital building, and, if not separated, a large system must operate to maintain preset conditions in a small space.

2.3.2 Laboratory Pressure Relationships

Good laboratory ventilation requires a careful balance between discharge and supply air volumes as well as careful placement of discharge and supply air locations. Even within the same type of laboratory, requirements may vary depending on the hazards of the materials being used, the quantity of hazardous materials being handled, and the nature of the operations involved. Communication with laboratory users at an early stage in planning helps to identify potential hazards before the final design stage. Laboratories using hazardous materials must be maintained at a negative pressure relative to hallways and other adjacent public access areas. Special purpose laboratories requiring positive pressure with respect to hallways and other adjacent areas will be discussed later. All offices, conference rooms, lunch rooms, and other public areas must be maintained at a slightly positive pressure relative to the laboratories to ensure safety.

2.3.3 Laboratory Ventilation Systems

There are three main types of ventilation systems for laboratories based on function:

1. *Comfort ventilation* to supply and remove air for breathing and maintaining design temperature and humidity, in part or in whole.
2. *Exhaust ventilation* that is designed specifically for health and safety protection.
3. *Makeup air* to replace the volume discharged to the atmosphere through the health and safety exhaust ventilation systems.

The three systems are described in detail in the following paragraphs.

2.3.3.1 Comfort Ventilation Supply Air for Laboratory Modules

All air exhausted from laboratories must be replaced with mechanically supplied air or by infiltration from adjacent areas. An adequate supply of air is essential to provide the correct number of room air changes needed to establish safe working conditions.

2.3.3.1.1 Supply Air Velocity and Entry Locations Inside Laboratories

Makeup air in equal quantities must be supplied to laboratories when air is exhausted through health and safety protection systems. Most of the makeup air will come in as air supplied directly into the laboratory but a small amount will infiltrate from adjacent spaces when they are at higher pressure. The location of makeup air outlets and the temperature of the air supplied are important. High-velocity supply air jets create excessive turbulence at a hood face that can disrupt exhaust ventilation system performance and be discomforting to occupants of the laboratory. It is recommended that high-velocity outlets be omitted in laboratory design.

Supply air grilles and diffuser should be selected and located so that the air velocity at the occupant's level does not exceed 50 ft/min.

2.3.3.2 Recirculation of Laboratory Room Air

All hazardous materials should be used in chemical fume hoods or with some other type of local exhaust ventilation. When the amount of exhaust air is reduced to a minimum by the use of well designed and well functioning chemical fume hoods and alternative local exhaust systems, less energy is expended and comfort is equal to what would be experienced were fumes dissipated by dilution air exchange involving the entire laboratory room. All air from local exhaust facilities, as well as from lavatories, sterilization rooms, and similar facilities, must be exhausted to the outside with no recirculation allowed. Air from offices, libraries, conference rooms, and similar nonlaboratory facilities can be recirculated with the addition of the minimum amount of outside air required to maintain health and safety and to comply with building codes.

2.3.4 Exhaust Ventilation for Laboratory Modules

Exhaust air systems of three types may be needed in each laboratory module: (1) removal of general room comfort supply air and contamination dilution ventilation air; (2) health and safety exhaust ventilation air from laboratory chemical fume hoods; and (3) local or spot exhaust ventilation air. The choice of system(s) will depend on the size of the laboratory, the nature and quantity of materials used, and the type of installed laboratory equipment. General room exhaust air is discussed in Section 2.3.4.1.

2.3.4.1 Exhaust of General Room Ventilation Air from Laboratories

The manner in which general ventilation air is exhausted from each laboratory room depends on its size and the nature of the activities and equipment present. In some cases, adequate exhaust of general room ventilation air will be provided by a laboratory fume hood or some other local exhaust air system. In other cases, a combination of general room return air facilities, chemical fume hoods, and additional local exhaust air facilities may be used.

All occupied and unoccupied laboratories require a minimum mechanical ventilation rate of 0.5 cubic foot per minute (cfm) per square foot of floor area when the fume hood is not operational. When a fume hood is operational, removal of a minimum of 1 cfm per square foot or the equivalent of six complete laboratory air changes per hour will be needed. It must be kept in mind that these figures represent the minimum exhaust air rates for laboratories. In most cases, larger air exhaust rates will be needed to handle all the air discharged to the atmosphere through laboratory hoods and other local exhaust air facilities inside the laboratory. Opportunities for energy conservation are discussed in Chapter 21.

2.3.4.1.1 Velocities for Removing Room Ventilation Air

Exhaust grilles for general room ventilation should be sized so that the inflow face velocity is between 500 and 750 fpm (feet per minute). Wall-mounted grilles should be placed to provide an airflow direction within the laboratory from the entrance door toward the rear of the laboratory to minimize the escape of fumes to the corridor.

2.3.4.2 Air Rates for Laboratory Hoods and Other Local Exhaust Air Facilities

Because air exhausted from these facilities cannot be recirculated and must be discharged to the atmosphere, the energy cost associated with their use is high. Opportunities for energy conservation are outlined in Chapter 21.

2.3.4.3 Chemical Fume Hoods

Laboratory hoods, sometimes called chemical fume hoods or fume cupboards, are a form of local exhaust ventilation commonly found in laboratories using toxic, corrosive, flammable, or malodorous substances. The purpose of a laboratory fume hood is to prevent or minimize the escape of contaminants from the hood to the laboratory air and by this means to provide containment. Successful performance depends on an adequate and uniform velocity of air moving through the hood face, commonly referred to as the control velocity. Hood performance is adversely affected by high-velocity drafts across the face, large thermal loads inside, bulky equipment in the hood that obstructs the exhaust slots at the rear, and poor operating procedures on the part of personnel using the hood by their failing to work 8 in. or more inside the hood face. With the sash closed, the hood can minimize the effects of small explosions, fires, and similar events that may occur within, but it should not be depended upon to contain fires or explosions other than trivial ones. To function correctly, a chemical fume hood must be designed, installed, and operated according to well-established criteria.

The chemical fume hood is the laboratory worker's all-purpose safety device. It is probably the single item that most definitively characterizes a laboratory and its importance for the safety and health protection of laboratory workers cannot be overstated. This being so, it is essential for the laboratory designer to understand thoroughly the functions that characterize a satisfactory laboratory hood and the several designs that are on the market. Not all are equally effective or efficient in the utilization of airflow.

Well designed fume hoods have several important characteristics in common: (1) air velocity will be uniform, i.e., ±20% of the average velocity over the entire work access opening; (2) all the hood surfaces surrounding the work access opening will be smooth, rounded, and tapered in the direction of airflow to minimize air turbulence at the hood face; (3) average face velocity will be a minimum of 100 fpm for work with any of the chemical, biological, and radioactive materials usually encountered in university, government, and industrial research, and commercial consulting laboratories. When handling substances associated with a somewhat higher hazard level, 125 fpm average face velocity is recommended. OSHA calls for but does not require average face velocities in excess of 150 fpm, for laboratory hoods used with any of the 13 carcinogens listed in OSHA 1910.1003 et seq. (OSHA, 1978). Hood face velocities in excess of 125 fpm are not recommended because they cause disruptive air turbulence at the perimeter of the hood opening and in the wake of objects placed inside the work area of the hood (Chamberlin, 1982; Ivany, 1986). (4) The face velocity of hoods with adjustable front sash will be maintained constant (within reasonable limits) by an inflow air bypass that proportions the air volume rate entering the open face to the open area, or by some other method that produces a similar result.

Use of a totally enclosed and ventilated glove box is recommended for handling very hazardous materials. The use of glove boxes minimizes air volume and simplifies air treatment for environmental protection.

There are a number of distinctive types of laboratory hoods in widespread use. Each will be identified in the following sections and described more completely in Appendix V, which includes purchase specifications.

For safety, especially when working with hazardous materials that give no sensory warning by odor, visibility, or prompt mucous membrane irritation of their escape into the workroom from the laboratory hood, it is advisable to provide each hood with an airflow monitor capable of giving an easily observed visual display of functional status. Installation of a hood airflow gage at the site of use has the special advantage of placing responsibility for day-by-day monitoring of hood function with the primary users. Devices that may be used to monitor hood function include liquid-filled draft gauges, Magnehelic gauges,* and bridled-vane gauges. A hood monitoring and airflow control system that has important energy conservation aspects is described in Chapter 21.

---

* Dwyer Instruments Inc., P.O. Box 373, Michigan City, IN 46360

2.3.4.3.1 Standard Chemical Fume Hoods

The basic chemical fume hood incorporates the four principles enumerated in Section 2.3.4.3 that characterize a well designed fume hood. Most laboratory furniture suppliers have one or more models that will meet the listed criteria. A model purchasing specification is presented in Appendix V.

Auxiliary air hoods are a basic type of fume hood that is acceptable for laboratory use. The major difference between the standard chemical fume hood and the auxiliary air hood is the method employed to provide makeup air to the hood. For standard hoods, all of the makeup air is provided by the room HVAC system, whereas, for the second type, sometimes called "supply air hoods," part of the air is introduced from a supply air grille just above and exterior to the hood face. Some manufacturers recommend up to 70% of the total hood exhaust volume may be supplied in this manner with the remainder coming from the laboratory HVAC system. When auxiliary supply air is introduced behind the sash (an incorrect hood design), the hood chamber is likely to become pressurized and blow toxic contaminants out the open front into the laboratory. The only kinds of auxiliary air hoods that are acceptable are those that introduce the auxiliary air exterior to and above the hood face.

Auxiliary air hoods have two advantages. First, the air supplied to the hood does not have to be cooled in warm climates, a significant energy saving. Second, auxiliary air hoods can be used in laboratories that contain so many hoods that the volume of supply air, were they all standard chemical fume hoods, would result in many more room air changes per hour than are desirable. Providing auxiliary supply air, even to only some of the hoods, reduces room air velocities and supply air quantities to more acceptable levels.

There are disadvantages to the use of the auxiliary supply air hoods. First, they are more complex in design than the usual chemical fume hood and their correct installation is more critical to safe and efficient operation. Often only a small imbalance between room air and auxiliary air supplies can result in unsafe operating conditions. Second, unsafe conditions occur when the velocity of external auxiliary air supplied to the face of the hood is excessive. High-velocity air sweeping down across an open hood face can produce a vacuum effect and draw toxic contaminants out of the hood. Third, this type of hood must have two mechanical systems (separate exhaust and supply systems) for each hood. Inasmuch as some supply air will need to be added to the laboratory anyway, the auxiliary air hoods require two supply air systems instead of the one supply system that standard chemical fume hoods require.

When the advantages outweigh the disadvantages, auxiliary air fume hoods should be used because energy savings through reductions in summer cooling and winter heating can be realized even in moderate temperature climates. Energy savings during the heating season occur because the auxiliary air need not be heated to as high a temperature as the comfort ventilation air. Use of auxiliary air hoods is particularly attractive for renovations when one or more fume hoods must be added to a laboratory with limited amounts of supply air capacity. Because of the complexity of this type of hood, the performance specifications are more stringent than for standard chemical fume hoods. Specifications for both types are contained in Appendix V. It is extremely important to choose the particular auxiliary air hood that will be purchased before the supply air system is designed because not all auxiliary air hoods have the same requirements for room and auxiliary air volumes and some meet performance requirements at less than 70% auxiliary air.

2.3.4.3.2 Horizontal Sliding Sash Hoods
Economy in the utilization of conditioned air for laboratory hoods can be achieved most satisfactorily by maintaining the required face velocity but restricting the open area of the hood face. A transparent horizontal sliding sash arrangement can cut overall air requirements by 50% if two half-width panels are used. Similarly, if three panels are used, the maximum open area reduction is 67% for a two-track setup and 33% for a three-track arrangement. This design also has an advantage over the conventional vertical sash because the full height of the hood opening is available. If panels 14 to 16 in. wide are used, they can also serve as safety shields.

Under most conditions, the conventional fume hood with horizontal sliding sashes gives personnel protection equal to a hood with a vertical sash provided it is designed and operated to have a minimum average face velocity of 100 fpm at the maximum face opening. All velocities in the plane of the hood face should be greater than 80 but less than 125 fpm under operating conditions. Sometimes, turbulence occurs at sharp panel edges when inflow velocities exceed design values. Visual smoke trails may be used to test for inward airflow across the entire face opening and an absence of turbulence at sash edges.

2.3.4.3.3 Perchloric Acid Hoods
Perchloric acid hoods require special construction, construction materials, and internal water-wash capability. Problems reported with hoods heavily used for perchloric acid digestions are associated with the consolidation of explosive organic perchlorate vapors that condense while passing through the hood exhaust system. They can detonate upon contact during cleaning, modification, or repair. Therefore, use of specially de-

signed fume hoods is required for use with perchloric acid. Perchloric acid hoods should meet the same contaminant fume retention capabilities as the chemical fume hoods described in the two preceding sections. In addition, they should be constructed of stainless steel and have welded seams throughout. No taped seams or joints and no putties or sealers can be used in the fabrication of the entire hood and duct system. The perchloric acid hood also requires an internal water-wash system to eliminate the buildup of perchlorates. The water-wash system should consist of a water spray head located in the rear discharge plenum of the hood plus as many more as are needed to ensure a complete wash down of all surfaces of the duct work from the hood work surface to the discharge stack on the roof. To drain all of the water to the sewer in a satisfactory way during normal wash-down operations, it is important to have the exhaust duct go straight up through the building with no horizontal runs. A straight vertical duct run also facilitates periodic examination for maintenance purposes.

2.3.4.3.4 Biological Safety Cabinets

The biological safety cabinet is a special form of containment equipment that will be addressed in more detail in Chapter 12 and Appendix Vf. It is used for work with cell and tissue cultures and parenteral drugs when the materials being handled must be maintained in a sterile environment and the operator must be protected from toxic chemicals and infective biological agents. The dual functions of protecting the worker and maintaining sterility are achieved by two separate cabinet flows: a turbulence-free downward flow of sterile HEPA-filtered air inside the cabinet for work protection, and an inward flow of laboratory air through the work opening to provide worker protection. In addition, all air exhausted from biological safety cabinets is filtered through HEPA filters to provide environmental protection. The inward airflow and the downward airflow are delicately balanced in the biological safety cabinet and great care must be exercised to maintain the design flow rate of each, as well as the ratio between the two. Biological safety cabinets have achieved widespread use in the last decade for recombinant DNA research.

2.3.4.3.5 High-Velocity–Low-Volume Spot Exhaust Systems

Effective exhaust ventilation must be provided for all apparatus and procedures used in the laboratory that generate hazardous contaminants or create excessive heat. In some cases it may not be possible or desirable to conduct the operations in a chemical fume hood. For example, if the equipment is large, it may not fit into a hood, or it may fit but occupy too much hood space and affect hood performance adversely. These cases call for special or supplementary ventilation arrangements that often take the form of high-velocity–low-volume local exhaust points consisting of open-ended exhaust hoses or ducts. Flexible exhaust ducts of 4 to 6 in. diameter, often referred to as "sucker hoses" or "elephant trunks," are useful for this service because they can be moved to locations where they are needed. An advantage of local exhaust hoses is their ability to reduce the total amount of air removed from the laboratory as a result of capturing contaminants at the source at high velocities, thereby using less total air volume than would be required for a fume hood. An ASHRAE study indicates that local exhaust systems are much more energy conservative than dilution ventilation (DeRoos, 1979).

Spot exhaust facilities require a high static pressure exhaust system that must be provided to the laboratory from a system separate from that serving the hoods. This is because hoods are low static pressure devices whereas spot exhaust points require from 2 to 5 inches of water gage (in. w.g.), depending on design factors such as the quantity of air and air velocity at the opening. It is essential to design for adequate local exhaust quantities in the planning stages because it is almost impossible to upgrade the system after installation except by total replacement.

2.3.5 Exhaust Fans and Blowers

After total air volume and static pressure requirements have been established, fan selection for a laboratory fume hood or spot exhaust system will depend upon the following factors: (1) ability to fulfill application requirements, (2) ease of maintenance, (3) initial cost, (4) life expectance, and (5) availability of spare parts.

Cast-iron fans are ideal for all exhaust air applications, including spot exhausts, because they are of excellent quality and reliability, have an extended lifespan, and require very little maintenance. Whenever long life expectancy and freedom from maintenance are not critical, steel plate fans are acceptable. Whenever the effluent air from fume hoods customarily contains large amounts of severely corrosive gases plus condensing water vapor, exhaust fans constructed of fiberglass-reinforced polyester (FRP) are highly recommended. Fume hood exhaust fans handling contaminants composed of dusts and mists, or containing flammable, toxic, or corrosive materials, should be located outside occupied areas of the building and as close as possible to the point of discharge to the atmosphere, preferably on the roof of the laboratory building.

Exhaust fans should have a discharge velocity at least 2500 fpm and a stack extending at least 10 ft above the roof parapet and other prominent roof structures. Under no circumstances should weather caps be used on local exhaust system stacks. To take care of precipitation into the open

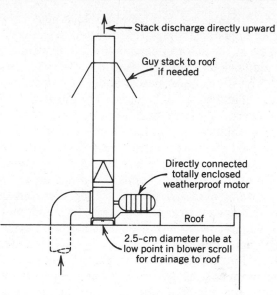

FIGURE 2-6. Exhaust stack and blower. From M. W. First, *Control of Systems, Processes, and Operations,* 3rd ed., Vol. 4, *Air Pollution* (A. C. Stern, Ed.), Chap. 1, Academic Press, New York, 1977.

stack when the fan is idle, there should be a drain connection at the low point of the scroll casing that can drain directly to the roof. For small blower-motor sets, there is a distinct advantage associated with selecting a direct-drive blower with a totally enclosed, weatherproof motor of the correct rpm. This selection eliminates belts, belt guard, and motor enclosure, and results in a compact, maintenance-free installation. A typical arrangement of a roof-mounted motor blower and stack is shown in Figure 2-6.

Belted fans should be double-belt driven, the shaft bearings should contain standard grease fittings for lubrication, and the fan should have a drain at the low point of the scroll casing. Vibration isolators are required to minimize noise transmission through the connecting ducts and building structure. The entire installation exposed to the weather must be built to withstand wind loads of 30 lb/ft^2 applied to any exposed surface of the fan and ducts rigidly attached to it.

Explosion-proof motors and nonsparking wheels are required when it is possible for the effluent air to contain more than 25% of the lower explosive limit (LEL) of any combination of vapors and aerosols. All fan motors must meet the requirements of the National Electrical Code (NFPA 70, 1985) and conform to applicable standards for load, duty, voltage, phase, frequency, service, and location. Motors should be mounted on an adjustable sliding base. Motors of 0.5 horsepower and larger should be of the squirrel-cage induction or wound rotor induction type and should have ball or roller bearings with pressure grease lubrication fittings. Drives for belted motors should be as short as possible and equipped with a matched set of belts rated at 150% of capacity. A weatherproof metal guard with angle iron frame securely fastened to fan housing and fan base should be provided for protection.

Each fume hood should be exhausted by its own exhaust fan. When multiple fume hoods are manifolded to a single exhaust fan, imbalances in the exhaust airflow can occur as hoods are turned on and off. This can be overcome by turning all hoods on and off together, but this is no longer considered a satisfactory arrangement from the standpoint of energy efficiency. Following installation of a manifolded hood system, the correct airflow to each hood must be established by adjusting dampers placed in the exhaust duct for that purpose and locking them in position. Balance is difficult to maintain and the system is prone to lose balance, making one or more hoods potentially unsafe. Cross-contamination from various hoods can also occur when the manifolded system is no longer in balance. Because of this and other potential problems, multiple fume hoods on a single exhaust fan are not recommended.

2.3.5.1 Exhaust Air Cleaning for Laboratory Effluent Air

Generally, exhaust air from laboratories is not cleaned prior to release because of the excellent dilution capability of the atmosphere when the air is discharged straight upward, starting 10 ft or more above all roof obstructions. However, certain laboratory procedures may require special effluent gas treatment to avoid polluting the atmosphere. Specific instances will be described in Part II, where detailed descriptions of commonly used laboratories will be found.

2.3.5.2 Exhaust Ducts and Plenums

2.3.5.2.1 Construction of Exhaust Ducts and Plenums

Exhaust ducts for fume hoods and local exhaust systems should, preferably, be of the high-velocity type to avoid settlement of particles in horizontal runs. Additionally, high-velocity systems reduce duct cross section and save space in vertical chases and above-ceiling utility areas. Stainless steel of a high chromium and nickel content is the material of construction for hard service in a corrosive, erosive, and high-temperature environment, but cost is high. Other materials of construction, such as epoxy-coated steel, can be used for less severe applications. Plastic piping has exceptional resistance to many commonly used corrosive chemicals but physical strength and flame spread ratings are less than for metals. It may be excessive but California regulations require a 2 h rated separation between fume hood exhaust ducts from different floors within the same chase. All fume hoods and local exhaust system ducts should be constructed of round piping with the interior of all ducts smooth and free of obstructions. All joints should be welded or otherwise sealed airtight.

Flexible ducts used for spot exhaust service should be kept to minimum lengths because flexible duct has an airflow resistance that is as much as two to three times that of metal duct of the same diameter. Poorly formed flexible elbows can increase system resistance to the point where function becomes lost. To reduce energy loss and air noise levels, the open ends of sucker hoses should be tightly capped when not in use. Flexible tubing must be selected from among the noncollapsible types.

2.3.5.2.2 Duct Leakage

All local exhaust system ductwork located inside the laboratory building must be maintained under negative pressure to prevent leakage of contaminated air into occupied spaces. Maximum duct leakage of 2% at design negative pressure should be specified in the building design documents and the installation carefully supervised. To ensure that this standard is achieved, the ducts should be tested, without fail, after installation by capping the ends of duct runs, putting the ducts under the design negative pressure specified in the construction documents, and measuring the volume of inflow air carefully with a sensitive, variable head airflow meter. Inflow rate should not be greater than 2% of the maximum design airflow rate for the duct run under test. It is necessary to point out that balancing measured airflow rates into a system against measured airflow at the discharge end is not an acceptable way of assuring no more than 2% inleakage because field measurements cannot be relied upon to less than ±5%, at best. If condensation inside the ducts is possible, all sections must be pitched to drain to a sewer connection.

Ideally, ductwork should be designed to be self-balancing without the use of trimming dampers. Dampers inside exhaust air systems serving laboratories (with the exception of those required for emergency fire suppression purposes) are failure-prone from corrosion, vibration, and maladjustment by unknowledgeable service personnel. Maladjusted dampers are frequently observed to seriously compromise the safety objective of the exhaust systems in which they are installed. Therefore, experienced ventilation engineers design systems that will meet building and laboratory design objectives without trimming dampers. To be effective, however, installation contractors must follow directions with sufficient exactitude to accomplish the desired results. Meticulous testing of completed systems is the only way to assure quality installations. However, the future flexibility of any installation may be compromised without dampers.

2.3.5.2.3 Noise Suppression

Air noise in laboratory ductwork can be minimized by keeping velocities within acceptable ranges, 2000–3000 fpm, and by using flexible connections between fans and ducts. Use of sound attenuators in exhaust ductwork is not an acceptable solution to noise problems, although their use is acceptable in supply air systems. Control of noise from sucker hoses (a potential major noise source) was covered in Section 2.3.5.2.1.

2.4 LOSS PREVENTION AND OCCUPATIONAL SAFETY AND HEALTH PROTECTION

2.4.1 Emergency Considerations

The clear objective of laboratory building design is to avoid and prevent unsafe conditions. In spite of the best planning, experience indicates that emergencies sometimes occur from unforeseen events. Therefore, the prudent building designer takes all reasonable steps to contain possible emergencies by the installation of loss control services and equipment. With emergency systems, personnel and physical plant losses can be avoided or at least reduced to a tolerable level. All of the following emergency considerations should be included in the basic design. Rejection of any one must be documented and justified. Those not required by law or code represent good practice standards.

2.4.1.1 Emergency Fuel Gas Shutoff

In facilities where fuel gas is piped throughout the building to laboratories, a method of shutting off the flow of gas at the laboratory must be installed for emergency use. The preferred method is to run the gas supply pipe to a wall just outside the laboratory and locate an emergency shutoff valve station at that point. The station should consist of a simple ball valve located in a box with a breakable glass or easily removable cover and a clear sign announcing its function. Although a major building gas shutoff must be provided with an excess-flow check valve to serve the entire building, the major building cutoff valve may be too remote from the laboratory room involved in an emergency and the building excess-flow check valve will be too large to sense and to stop the flow of fuel into small laboratory areas. For both reasons, local valve stations are necessary.

Note. The requirements for fuel gases do not apply to nonfuel laboratory gases that are piped into laboratories through small piping systems from a central source. However, nonfuel gas pipes, valves, excess-flow check valves, and other materials and installations must meet applicable codes and standards of the NFPA, the Compressed Gas Association, and ANSI.

2.4.1.2 Ground Fault Circuit Interrupters

Ground fault circuit interrupters should be used on all laboratory benches and where portable or nonstationary equipment is used. Ground fault circuit interrupters are devices that compare the current flow in wires feeding and returning from electrical devices, such as mixers, ovens, meters, blenders, pumps, and stirrers. When an imbalance of more than 5 to 8 mA occurs, the electrical power to the device will cut off. This is predicated on the assumption that the leakage, or lost current, could pass through the body of an operator. The circuit interrupter functions to limit the amount of "shock" to nonlethal levels. This is in contrast to the electrical fuses and circuit breakers normally found in a laboratory building. They do not open until the power requirements of the equipment on-line are exceeded and, therefore, fuses and circuit breakers cannot prevent an electrical shock to personnel; they function only to protect equipment and the building against fire.

Ground fault circuit interrupters tend to sum up all the leakage of devices plugged into one circuit. Installation of more interrupters, or fewer electrical appliances per circuit, can help to reduce nuisance tripping. Old electrical devices or appliances, whose insulation is not as good as when they were new, also can trip the interrupters, indicating a need for these appliances to be renovated or replaced. Fuses and circuit breakers, ground fault circuit interrupters, and a comprehensive electrical safety program are all necessary if maximum freedom from shock, electrical fire, and adverse electrical equipment problems is to be maintained.

Ground fault circuit interrupters should be used at all laboratory benches and there should be no more than three duplex outlets per interrupter. Ground fault circuit interrupters should also be used near wet operations such as sinks. Locating the ground fault circuit interrupter unit in a central box at the main electrical panel for the laboratory encourages prompt attention to standing line leaks and other potential problems in the system. The use of no more than three duplex outlets placed close to each other and attached to a ground fault circuit interrupter on the same bench is ideal because this arrangement minimizes long wire runs with many junction boxes that have a leak potential. In addition, when ground faults occur that shut the system down, they can be corrected easily and the system turned back on again quickly.

Stationary electrical equipment, such as refrigerators and ovens, should be equipment-grounded through hard wiring; therefore, these devices need not be ground fault circuit interrupted in addition. One duplex plug outlet on each laboratory bench may be installed without ground fault circuit interrupters to provide a means of temporarily using equipment that would otherwise break the circuit. This outlet should be well identified.

2.4.1.3 Master Electrical Disconnect Switch

In each laboratory area, electrical services should be identified and sited in such a manner that all electrical power to the laboratory (except lighting and other life-safety critical items such as exhaust systems and alarms) can be quickly disconnected from one easily accessible location. If such an arrangement is not possible or feasible, a shunt trip breaker system should be installed that will accomplish the same thing upon depression of the master mushroom kill-switch. This switch should be located near the normal emergency exit route, but not in a place where it can be activated inadvertently.

2.4.1.4 Emergency Showers

Emergency deluge showers are used to dilute and wash off chemical spills on the human body. Because many chemicals attack the body rapidly, the location and reliable functioning of the showers is of critical importance. Specifications for emergency showers are in Appendix VII.

Deluge showers, when properly installed, provide a minimum of 30 gal of water per minute and deliver this volume at low velocity because high-velocity showers can further damage injured tissue. For this reason, only low velocity deluge shower heads should be used. The valve operating the shower should be of the type that requires a positive action to close, such as a ball valve with a non-spring-loaded level arm. Rigid pull bars of stainless steel stand up better under corrosive conditions than other metals. When testing is not performed routinely, the valves may become so difficult to operate that it could result in a chain pull breaking before the water valve is turned on. The preferred location of a shower is just outside the hazardous area. This could be in a hall not more than 25 to 50 ft from the laboratory. The valve should be located close enough to a wall that normal traffic will not bump into the operating rod but far enough out from the wall to allow a second person to move around and help an injured person. A large contrasting spot should be painted, embedded in, or affixed to the floor directly beneath the shower to indicate its location.

At least one shower in each area of the building should be tempered to provide water at 70 to 90°F to accommodate an injured person who should remain in the shower for the recommended 15-min period. Normally, cold water temperature is substantially below 70°F and immersion in water at this temperature would be painful for an injured person. The remainder of the deluge showers can be connected directly to the potable cold water system. Another method of providing a tempered water shower would be to use an existing toilet facility with a shower stall already in place. In this case, a standard emergency shower deluge head could be installed with the correct pipe size. With this arrangement, the injured person would be able to have a degree of privacy for the 15-min shower, especially important if it became necessary to remove clothing.

Another advantage of using a shower stall in a rest room for a tempered emergency shower is that the runoff water can go through a floor drain. Floor drains are not normally installed in corridors under emergency showers because they are used so infrequently that the traps become dry and allow sewer gases to enter the area. If the tempered water shower is to be in a hall corridor or laboratory, a floor drain capable of handling the entire output of the shower should be installed.

Contrary to popular opinion, emergency deluge showers are not installed to extinguish clothing fires. The best method of extinguishing a clothing fire is to "stop, drop, and roll," and then remove the burned clothing and seek help. Nevertheless, if one is within a step or two of a deluge shower, this device can be used as an effective fire extinguisher for a clothing fire. The hazard associated with moving from the laboratory to a shower with clothes burning is the probability of inhaling burning gases and searing the breathing passages, including the deep parts of the lungs, to a fatal degree. The "stop, drop, and roll" method is preferred.

2.4.1.5 Emergency Eyewash

The reaction of many chemicals with the human eye is very rapid. For example, the interior chamber of the eye can be reached within 6 to 8 s following a splash of concentrated ammonia. Therefore, most physicians agree that an immediate flush with copious quantitites of water is the best first aid treatment for chemical splashes to the face and eyes.

There are two basic types of emergency eyewash devices available: plumbed and portable. Because many portable units do not have the capacity to deliver 15 min of copious flushing, this discussion is limited to plumbed emergency eyewash units. It is acknowledged that portable units have value when they can be located very close to the user and result in a very quick start to eye flushing that can be completed at the plumbed eyewash station once the initial flush has been accomplished.

There should be at least one eyewash facility per laboratory. They may be located at sinks or at any other readily accessible area in the laboratory. Laboratories using strong acids or bases should have an eyewash within 20 ft of the hazard area. A tempered water unit should be installed for each contiguous group of laboratories. It should be set at 70 ± 5°F. This is necessary because most physicians recommend a 15-min wash prior to transportation to a medical facility. Holding one's eyes open to 35–40°F water for more than a minute or two can become painful and ultimately impossible.

The best units for washing eyes and face are the multistream, cross-flow types. They flush the face and both eyes at the same time with near zero-velocity water. Hand-held units on a hose, such as a kitchen spray, have the advantage of serving as a minishower for splashes of the arms, hands, and other small spills to the body. Specifications for emergency eyewash facilities appear in Appendix VIII. See also, A. Weaver and K. Britt, *"Criteria for Effective Eyewashes, and Safety Showers"* (ASSE, 1977).

2.4.1.6 Chemical Spill Control

A means for effecting prompt spill control should be provided for all laboratories using hazardous chemicals. They include making provisions for (1) establishment of convenient cleanup stations; (2) one or more storage locations of adequate size to hold the requisite quantity of neutralizing chemicals, adsorbents, and other equipment needed to deal effectively and rapidly with the maximum anticipated spill; and (3) diking materials, consisting of fixed dikes or dike bags. The locations of cleanup stations need to be clearly identified in the building design stages.

2.4.1.7 Emergency Cabinet

Emergency cabinets should be provided for all laboratories and located in an area that will be readily accessible under stress conditions. The cabinet should be sized to hold items specific to the work of the laboratory, as well as a number of general items. General and specific items may include:

General	Specific
Emergency blanket	Medical antidotes
Emergency response information	Chemical spill kits
First aid kit	Protective clothing
Stretcher	Escape breathing equipment
Resuscitation equipment	

2.4.2 Construction Methods and Materials

The fire-resistive construction requirements of BOCA Article 9 for Class A buildings should be followed (BOCA, 1985). Selection of wall coverings, bench materials, and other furnishings of some laboratories will require special consideration. Examples are installation of materials resistant to fire, consideration of the potential for electric shock in the use of metal furnishings, reflectivity of building materials for certain operations involving light, darkness, and lasers.

Adequate electrical outlets for standard line voltage equipment should be provided within each laboratory to facilitate the use of all pieces of anticipated electrical equipment. Older laboratories often have many extension wires draped around the laboratory, resulting in an unsafe condition. This has occurred because laboratory designers of a former period could not envision the enormous increase in the number of electrically driven devices in all modern laboratories. Unsafe extension wire conditions can be avoided in new laboratories by adequate preplanning that includes a systematic and realistic evaluation of current and future electrical outlet requirements. Electrical outlets in laboratories should be equipped with ground fault circuit interrupters (Section 2.4.1.2).

2.4.3 Control Systems

Equipment using hazardous materials, such as constantly flowing toxic or flammable gases, as well as electricity-dependent operations, should have alarm and automatic shutoff circuits capable of rendering the process and equipment safe in case of failure. To build these safeguards into a laboratory requires that potential problems be defined by the expected users of the laboratory during the building and laboratory design stages.

2.4.4 Alarm Systems for Experimental Equipment

A system of electronic communications should be installed in laboratory modules so that, when necessary, monitoring of key laboratory operations can be initiated to signal equipment malfunctions. The system should connect the laboratory to control points selected by the laboratory manager. Selected control points may be the office of the investigator, a hall outside the laboratory, a security guard station, or the fire station. The alarm system is not intended to be a control system; it should be for the transfer of information only. When there will be firm regulations against leaving experiments, reactions, and all other laboratory activities unattended, and when no other need exists, alarm systems for experimental equipment may be omitted. In an operation where hazards are controlled by outside services, such as water for a still, ventilation, or electricity, provisions should be made for shutdown of the operation in emergency situations. Emergency electrical systems are described in more detail in Section 2.4.1.3.

2.4.5 Hazardous Chemical Disposal

Discarded hazardous chemicals, including flammable liquids, strong reagents, and highly toxic chemicals, should be segregated into a specific area within the laboratory for disposal. The hazardous waste storage area should not be located where an unexpected reaction could immediately involve persons working at normal work stations, or block normal egress routes. In addition, the hazardous waste storage area should be chosen so that, should an undesired reaction occur, it would not affect other areas of the building.

2.4.5.1 Hazardous Chemical Storage

Hazardous wastes should be stored in appropriate safety containers. Flammable liquid wastes, for example, should be stored in approved waste safety cans that carry UL or FM (Factory Mutual) approval for this use. They should be placed in a location where spills can be caught and retained without damaging the floor or creating slippery work and walking surfaces. Local ventilation may be necessary for especially toxic, malodorous, and volatile waste materials. Because some waste containers are large and bulky, the architect and laboratory designer should consult manufacturer's catalogs to get a good idea of the space that will be needed and to make provision for movement of the accumulated wastes by cart or some other suitable means. For more on chemical storage, see Section 1.4.6.

2.4.5.2 Chemical Waste Treatment Prior to Disposal

Laboratory sinks and floor drains should be tied into a chemical process sewer for the disposal of low hazard chemicals requiring dilution, neutralization, or some other treatment to make them suitable for ultimate disposal in the sanitary sewer system. Local, city, and state regulations should be examined to determine the extent to which dilution and neutralization tanks must be used.

2.4.6 Chemical Storage and Handling

Determining chemical storage and handling locations and methods after a laboratory has been constructed often leads to an acceptance of an unnecessary risk. These matters should be considered during the design phase of the laboratories. Experience has proven that site inspection of currently used facilities of the potential users of new laboratories is the best way to determine the amounts of chemicals that must be stored in the new laboratory. If an inspection is not possible, the information obtained from the expected users should be increased by a safety factor of 1.5 to 2.0 to allow for growth, and space provided accordingly.

2.4.6.1 *Storage in the Laboratory*

For safety, the quantity of chemicals stored in a laboratory should be kept to a minimum at all times and they should be stored by methods and locations appropriate to their hazard classification. Large supplies of chemicals should be stored in a central storage area serving the entire laboratory complex (see Section 1.4.7). The laboratory director or P.I. should be consulted to determine maximum storage quantities for each laboratory, and space should be provided in the new facility for adequate chemical stocks based on the information obtained.

2.4.6.2 *Standard Reagents, Acids, and Bases*

Strong acids and bases may be stored in the ventilated base of chemical fume hoods but separation should be provided to prevent cross-mixing in the event of breakage or leakage. Mild acids and bases such as citric acid and sodium carbonate may be stored with other low-hazard reagents. Open shelves for chemicals should be located out of normally traveled routes and have a ¾-in. lip to prevent movement over the edge due to vibration.

2.4.6.3 *Flammable Liquid Storage*

Contrary to a commonly held belief, flammable liquid storage cabinets are intended to protect the contents from the heat and flames of an external fire rather than to confine burning liquids within. Flammable liquids should be kept in UL-approved flammable liquid storage cabinets. The cabinets should be remote from other operations within the laboratory that could become involved in a fire. In addition, they should not be located where they could impede access to an exit in case of fire. Whether or not flammable liquid storage cabinets should be ventilated depends on several factors and, at the time of this writing, opinions are mixed. When a flammable liquid cabinet must be located under a chemical fume hood (although this is not desirable), it should be provided with minimal ventilation by being connected to the hood exhaust system through a flash arrestor. Typically, an exhaust connection is made into the back of the hood with one or two 1 ½-in.-diameter pipes that extend from the under-hood storage space through the work surface and into the hood plenum. If a storage cabinet for flammable liquids is to be located in some other part of the laboratory, remote from a chemical fume hood, and it is feared that it may generate flammable vapors and malodors from spills, it should be ventilated at a rate of three to five air changes per hour. Another plan is simply to leave the vent ports open with the flash arrestors in place and let the room exhaust handle any vapors escaping from the cabinet. Each of the methods cited is legally acceptable. The concern over the use of forced ventilation inside the cabinet is a fear of overcoming the effect of the flash arrestors in protecting the contents of the cabinet from an exterior fire.

2.4.6.4 Special Chemicals

Especially hazardous chemicals, and chemicals requiring security control, should be stored in locked cabinets located within the laboratory or in a central chemical storage room. The locked storage cabinets should be adequate in size and contain internal separations to provide for the storage of incompatible chemicals such as perchloric acid and cyanide compounds. Lecture bottles and full-sized cylinders of compressed gases should be stored in a ventilated storage area. Mechanically ventilated hood bases, and other types of vented cabinets, are suitable for this purpose.

2.4.7 Compressed Gas Cylinder Racks

When large compressed gas cylinders are required inside a laboratory, special facilities must be provided for their safe use. They include strapping and anchoring devices, adequate room ventilation to remove leaking gas, and easy accessibility for periodic exchange of cylinders. The area should be large enough to accommodate an extra tank of each gas in use (or at least the most used gases) with adequate room for empties awaiting collection. The direction of ventilation airflow should be out of the room and building immediately after passing the gas storage area. Local piping systems for gas cylinders located in the laboratory should meet the same criteria cited in Section 1.4.8. Local spot ventilation can also be used for satisfying gas cylinder exhaust requirements.

2.4.8 Safety for Equipment

2.4.8.1 Safe Equipment Locations

All equipment installed inside and outside laboratories should be located in such a manner that its failure will not involve other pieces of vital equipment, block egress routes, or create situations that overwhelm the capabilities of the ventilation and sprinkler systems (see also Sections 1.3 and 1.4.4). Very heavy pieces of equipment must not be placed or moved through areas where floor loading capacity has not been determined by informed study to be adequate for the imposed load.

2.4.8.2 *Material Locations*

Secure locations should be established for materials that could become involved in creating or worsening emergency situations. They include trash and waste baskets, waste chemicals, stored chemicals, and normal combustibles, such as supplies of single-use disposable laboratory materials.

2.5 SPECIAL SERVICES

Special utility services, such as high electrical demand, high-pressure steam, and hydraulic systems, must be installed in accordance with applicable codes and standards when they are available. When experimental use of special services is not covered by standards or codes, a review team of experts will be required to generate local substitutes for nonexisting codes and standards. This is an important safety provision. The more the special laboratory services deviate from those normally provided to all laboratories, the greater is the safety need for special review of equipment and usage patterns. The review team should include members of the industrial safety and industrial hygiene professions, building maintenance personnel, the user, and the architect-engineer designing the building.

II

DESIGN GUIDELINES FOR A NUMBER OF COMMONLY USED LABORATORIES

The following chapters address safety and health issues for a number of well defined laboratory types. Laboratory type is determined largely by the hazardous properties and quantities of the materials and equipment normally used, the work activities performed, and any special requirements of the laboratory that may affect safety and health adversely inside or outside the laboratory. It is of extreme importance that a specific laboratory type be selected in collaboration with laboratory users plus safety and industrial hygiene advisors. After the laboratory type has been selected, all of the issues discussed in the chapter of this manual dealing with that special laboratory type should be evaluated and implemented in the design stages.

The items discussed under Common Elements of Laboratory Design (Part I) apply to all laboratories except when specifically excluded in the chapter on that specific laboratory type and then an alternative will be given. In each of the specific laboratory types, special requirements unique to that laboratory will be addressed. They may supplement or supercede the requirements of the Common Elements section. It is recognized that for renovation projects it may not be possible to comply with all requirements due to constraints imposed by the existing facility, but the intent of the safety and health recommendations should be observed carefully when making compromises, and safety and industrial hygiene personnel should be consulted for professional advice.

A matrix showing the major safety and health considerations that should be addressed for each laboratory type appears in Part IV (Appendix XI). It should be used as a design refresher and as a quick overview. It should not be used to replace the more detailed information contained in the text.

3

General Chemistry Laboratory

3.1 DESCRIPTION OF GENERAL CHEMISTRY LABORATORY

3.1.1 Introduction

A general chemistry laboratory is designed and constructed to provide a safe and efficient workplace for a wide variety of chemical activities associated with analysis, quality control, and general chemical experimentation. Preferably, this type of laboratory will be located in a building containing mainly similar laboratory units, but not necessarily doing general chemistry work. For example, these laboratories do not belong in office buildings.

3.1.2 Work Activities

General chemistry laboratory activities include mixing, blending, heating, cooling, distilling, evaporating, diluting, and reacting chemicals as part of testing, analyzing, and research experiments. Most of this work will be conducted on a bench or in a laboratory chemical fume hood.

3.1.3 Equipment and Materials Used

Stills, extraction apparatus, reactor vessels, heaters, furnaces, evaporators, and crystalizers are standard items of equipment found in general chemistry laboratories. Many analytical devices (e.g., atomic absorption spectrometers, gas chromatographs, spectrophotometers) are commonly used in these laboratories as auxiliary equipment. Typical equipment that should be considered for placement includes gas cylinders, ovens, stills, experiment frames, vibration-sensitive balances, chemicals, glassware, and cryogenics.

Hazardous materials used in general chemistry laboratories include small quantities of chemicals of high toxicity, volatile liquids, dusts, compressed gases, and flammables.

3.1.4 Exclusions

General chemistry laboratories are not designed for handling extremely hazardous chemicals or performing especially hazardous operations. Hazardous operations not recommended for general chemistry laboratories include, but are not limited to, the use of (1) carcinogenic, mutagenic, or teratogenic chemicals; (2) highly explosive materials in greater than milligram quantities; (3) high-tension voltage and high-current electrical services; (4) radiofrequency generators and all electrical operations with a high potential for fire, explosion, or electrocution; (5) lasers of over 3 mW output power with unshielded beams; (6) gas pressures exceeding 2500 psig; (7) liquid pressures exceeding 5000 psig; and (8) radioactive materials in greater than $1-\mu\text{Ci}$ amounts. The general chemistry laboratory usually needs no special access restrictions except when specialized or highly sensitive equipment and toxic materials are to be used. If such operations are to be carried out routinely, the designation of the laboratory as a general chemistry laboratory should be reevaluated.

3.2 LABORATORY LAYOUT

A suggested layout for a general chemistry laboratory is provided in Figure 3-1. All the items described in Sections 1.2 and 2.2 should be reviewed and those that are relevant should be implemented.

3.3 HEATING, VENTILATING, AND AIR CONDITIONING

All the items described in Sections 1.3 and 2.3 should be reviewed and those that are relevant should be implemented.

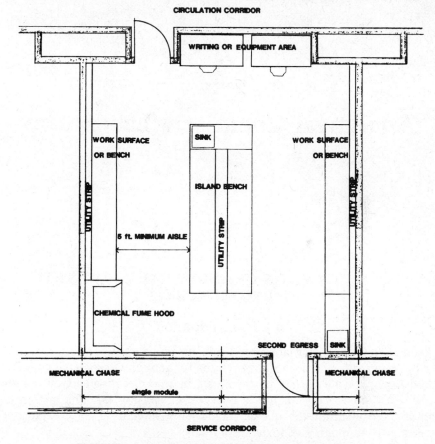

FIGURE 3-1. General chemistry laboratory: Sample layout.

3.4 LOSS PREVENTION, INDUSTRIAL HYGIENE, AND PERSONAL SAFETY

All the provisions in Sections 1.4 and 2.4 should be reviewed and implemented.

3.5 SPECIAL REQUIREMENTS

None.

4

Analytical Chemistry Laboratory

4.1 DESCRIPTION OF ANLYTICAL CHEMISTRY LABORATORY

4.1.1 Introduction

An analytical chemistry laboratory is designed, constructed, and operated to provide a safe and healthful work area for activities associated with the analysis of a wide variety of chemicals in amounts ranging from nanograms to kilos. Many analytical procedures call for handling moderate amounts of hazardous chemicals, including petroleum solvents, explosive gases, and all manner of toxic substances.

4.1.2 Work Activities

The activities performed in analytical chemistry laboratories include sample preparation involving mixing, blending, ashing, and digesting. The major activity is to analyze samples and this requires the operation of a variety of analytical instruments, some of which contain radioactive sources and many of which utilize hazardous radiations such as laser illumination, ultraviolet, infrared, and microwaves.

4.1.3 Equipment and Materials

The many kinds of analytical equipment utilized in this type of laboratory include spectrophotometers, gas and liquid chromatographs, mass spectrometer, balances, microscopes, stills, extraction apparatus, ovens, and furnaces. Heavy use of some of this equipment may generate a significant heat load in the laboratory.

Hazardous materials used in analytical chemistry laboratories include small quantities of chemicals of high toxicity, volatile liquids, dusts, compressed gases, and flammables. Toxic materials may be reacted or decomposed into nontoxic compounds during the analytical procedures, but usually, they remain in a toxic state during analytical manipulation. Occasionally, nontoxic components may react to produce hazardous reaction products, but this is not usual.

4.2 ANALYTICAL CHEMICAL LABORATORY LAYOUT

The layout of an analytical chemical laboratory is similar to that of a general chemistry laboratory (see Figure 3-1). Each item addressed in Sections 1.2 and 2.2 should be evaluated for its applicability to the specific needs of those who will use the analytical chemical facility and items that are relevant should be implemented.

Because it is likely that many pieces of analytical equipment of substantial size will be present in an analytical chemistry laboratory at all times, special care should be given to their location relative to egress routes, ventilation patterns, and the interactions of laboratory personnel with chemical handling operations. Safety and industrial hygiene personnel should be consulted for assistance with special problems of hazardous chemical handling and laboratory ventilation.

4.3 HEATING, VENTILATING, AND AIR CONDITIONING

4.3.1 Introduction

All the items described in Sections 1.3 and 2.3 should be reviewed and those that are relevant should be implemented.

4.3.2 Additional HVAC Needs

Special consideration should be given to providing good coverage of the laboratory with local exhaust systems. The effluent gases from some analytical devices, such as gas chromatographs and atomic absorption spectrometers, often contain toxic chemicals that need to be controlled at the source of generation by being vented to the out-of-doors. In some cases, instrument manufacturers provide recommendations specific to their instruments. Otherwise, the *Industrial Ventilation Manual* (ACGIH, 1986) or a qualified industrial hygienist should be consulted for advice. If perchloric acid use is anticipated, the safety measures described in Chapter 2.3.4.3.3 should be complied with.

The heat-producing capabilities of each of the analytical instruments and auxiliary equipment should be evaluated when estimating heating and air conditioning requirements.

4.4 LOSS PREVENTION, INDUSTRIAL HYGIENE, AND PERSONAL SAFETY

4.4.1 Introduction

All of the items described in Sections 1.4 and 2.4 should be evaluated for their applicability to the specific analytical laboratory under consideration and particular attention given to the location of analytical equipment in relation to egress routes and interaction with other equipment or chemical handling operations. Safety and industrial hygiene personnel should be consulted when additional advice is needed.

4.5 SPECIAL REQUIREMENTS

None.

5

High-Toxicity Laboratory

5.1 DESCRIPTION OF HIGH-TOXICITY LABORATORY

5.1.1 Introduction

A high-toxicity laboratory is designed and operated to provide safe use of highly toxic chemicals, including the use of carcinogens, mutagens, and teratogens. Industrial hygiene and safety personnel should be consulted for assistance in determining whether the nature and quantity of chemicals that will be used in a proposed laboratory fall into this category. Other sources of assistance include local OSHA and NIOSH offices and State Departments of Occupational Health and Industrial Hygiene. The National Institute of Health has published a "Suspected Carcinogens" listing and an internal guideline document entitled "Guideline for the Laboratory Use of Chemical Carcinogens" (NIH, 1981) that provides specific information on carcinogens. Not all of the design considerations included in this chapter may be needed in every high-toxicity laboratory. Close communication between all involved in the planning process will be necessary to determine specific requirements.

5.1.2 Work Activities

The basic experimental procedures used in high-toxicity laboratories are similar to those conducted in general chemistry and analytical chemistry laboratories, but provisions should be made for the additional safety procedures that will be required when handling highly toxic chemicals in more than microgram quantities. Although this section describes a laboratory that is similar to a general chemistry laboratory (Chapter 3), the design and operation of the safety provisions will be much more critical. The safety guidelines outlined here can be applied to other laboratory types such as chemical engineering and physics laboratories, whenever they use highly toxic materials in quantities that do not exclude their use in such laboratories.

5.1.3 Equipment and Materials Used

The equipment used in a high-toxicity laboratory will vary depending on the nature of the work, but, in general, the equipment will be similar to that found in a general chemistry or analytical chemistry laboratory.

5.1.4 Exclusions

Excluded from this chapter are the use of ionizing radiation, biological agents, and animals. Their use with high-toxicity chemicals requires additional design features that are addressed in Chapters 11, 12, and 18, respectively.

5.1.5 Special Requirements

The nature of the materials used make it necessary for the high-toxicity laboratory to have special access restrictions. OSHA has stringent requirements that must be followed for specific chemicals (OSHA, 1978). More recent OSHA regulations must be consulted to determine if important changes have been promulgated since the cited edition. In some cases, provisions must be made for change rooms and showers. Industrial hygiene and safety personnel should be consulted for advice when high-toxicity laboratories are to be built or existing laboratories converted to this use.

5.2 LABORATORY LAYOUT

5.2.1 Introduction

Many types of laboratory layouts are possible, depending on the specific nature of the work to be performed and the space available. A layout similar to a general chemistry laboratory (Chapter 3) is often adequate.

Specific work-space utilization layouts are described in *Safe Handling of Chemical Carcinogens, Mutagens, Teratogens, and Highly Toxic Substances* (Walters, 1980). Using either reference source as a starting point, all the recommendations for safety and health contained in Part I should be reviewed and those that are relevant should be implemented. In addition, the following items should be considered for their applicability to the laboratory work to be performed. Industrial hygiene and safety personnel, as well as appropriate state and federal agencies, may have to be consulted early in the planning phase because adoption of some of the recommendations contained in Part I may depend on the specific nature and quantity of the chemicals to be used in relation to applicable regulations for their use and safe disposal.

5.2.2 Change, Decontamination, and Shower Rooms

Adequate facilities must be provided for laboratory workers to change and shower because procedural requirements for working with highly toxic materials necessitate the use of frequent changes of protective clothing. For use of toxic materials some OSHA requirements specify that a shower must be included as an essential part of a high-toxicity laboratory; use of other toxic materials requires merely the availability of a shower in the building. Traffic flow into change and shower rooms should be designed so that there are separate clean and dirty pathways to and from the facility and there is no way to bypass the shower on the way out of the high-toxicity laboratory. (See Chapter 12 for more information on this topic.)

5.2.3 Work Surfaces

Work surfaces should be constructed from impervious and easily cleanable materials such as stainless steel. Strippable, epoxy-type paint is acceptable. Use of disposable bench coverings during work should be considered as an added safety practice.

5.2.4 Floors and Walls

Floor coverings should be of monolithic (seamless) construction and utilize materials such as vinyl or epoxy that are impervious to most chemicals and easily formed into seamless sheets. All cracks and construction seams in floors, walls, and ceilings should be sealed with epoxy or another chemically resistant, long-lived sealant. Utility conduits should be epoxy-sealed wherever they penetrate floors, ceilings, and walls. All laboratory lighting should be sealed with similar materials to be vaporproof and waterproof.

5.2.5 Handwashing Facilities

There should be readily accessible handwashing facilities located within the high-toxicity laboratory, as well as in change and shower rooms.

5.2.6 Access Restrictions

All entrances to a high-toxicity laboratory should be posted with permanent signs indicating restricted access due to the use of specific classes of chemicals (e.g., carcinogens, mutagens). The use of special key access should also be considered when a security breach is liable to result in serious illness or dispersal of high-toxicity materials outside the laboratory.

Isolation of the laboratory by tightly sealing doors, makes it desirable that there be a viewing window in each door. For the same reason, when flammable solvent usage is heavy, large blowout windows in the exterior walls are recommended. See NFPA 68 for the appropriate sizing calculations of the blowout areas.

5.3 HEATING, VENTILATING, AND AIR CONDITIONING

5.3.1 Introduction

All of the items described in Section 2.3 should be reviewed and those that are relevant should be implemented. Additional recommendations are given below. Industrial hygiene personnel should be consulted for assistance when safety and health situations not covered by this manual are encountered.

5.3.2 Laboratory Fume Hood

An average face velocity of 100 fpm is recommended. We do not recommend that it be increased to 150 cfm/ft^2 as an added safety feature for work with highly toxic materials because several investigators have indicated that it does not provide added protection (Chamberlin, 1978; Ivany, 1986). However, it should be noted in this connection that ACGIH and NIOSH do recommend a higher exhaust volume. Therefore, this recommendation should be tempered by reference to all current and applicable local, state, and federal regulations.

5.3.3 Glove Box

Some highly toxic materials require the use of a completely enclosed, exhaust-ventilated work space rather than a conventional laboratory fume hood. In these instances a glove box is required. It should meet the specifications defined by the American Conference of Governmental Industrial Hygienists (ACGIH, 1986), as outlined in Appendix Ve.

The glove box should be maintained as a closed system at all times and kept under a negative pressure of 0.25 in. w.g. It should be thoroughly decontaminated prior to exhaust airflow being shut down to avoid loss of its toxic contents to the laboratory.

5.3.4 Spot Exhaust for Instruments

Instruments used to weigh, manipulate, and analyze highly toxic chemicals should have spot exhaust ventilation at each potential source of contaminant release, or be completely enclosed in an exhaust-ventilated enclosure. Specific design requirements will vary with each type of equipment and chemical used. Consultation with the manufacturer of the equipment and an industrial hygiene engineer is recommended when the design and application of exhaust ventilation facilities is not obvious.

5.3.5 Storage Facilities

All facilities used for storage of highly toxic materials, such as cabinets and refrigerators, should be provided with exhaust ventilation to maintain airflow in an inward direction and prevent buildup of toxic contaminants within the storage space. Slot ventilation around refrigerator doors can be very effective.

5.3.6 Filtration of Exhaust Air

The air exhausted from fume hoods, glove boxes, and spot exhaust hoses should be decontaminated before release to the environment. The first cleaning stage should be a HEPA filter with a minimum efficiency of 99.97% for 0.3-μm particles when toxic aerosols are present. The second stage should be an activated charcoal adsorber when toxic vapors are present. The size will depend on the total quantity of airflow. All replaceable components should be capable of being changed without exposure of maintenance personnel (e.g., bag-in, bag-out procedures). For some chemicals, an adsorbent other than activated charcoal may be necessary or more desirable. In case of doubt, an industrial hygienist should be consulted.

5.3.7 Directional Airflow

Air infiltration should always flow from uncontaminated to contaminated areas, that is, from corridors to change and decontamination rooms, and, finally, to the high-toxicity laboratory, itself. Flow direction should be monitored by appropriate devices consisting of differential pressure sensors equipped with audible and visual alarms to warn of upset conditions.

5.4 LOSS PREVENTION, INDUSTRIAL HYGIENE, AND PERSONAL SAFETY

5.4.1 Introduction

All the items described in Section 2.4 should be reviewed and those that are relevant should be implemented.

5.4.2 Protection of Laboratory Vacuum Systems

Laboratory vacuum systems should be protected from contamination by installation of traps containing disposable HEPA and/or activated charcoal filter systems.

6

Pilot Plant (Chemical Engineering Laboratory)

6.1 DESCRIPTION

6.1.1 Introduction

A pilot plant, or chemical engineering laboratory, is designed, constructed, and operated to provide a safe and healthful work area for activities associated with the handling of substantial quantities of toxic chemicals, petroleum fuels, compressed gases, and other hazardous materials for chemical processing experimentation. A special characteristic of a pilot plant, in addition to large floor area and multistory height, is that materials are usually handled in large quantities (gallons and pounds) as opposed to the small quantities (milliliters and grams) used in most other types of laboratories. Because of the frequent use of flammable and explosive chemicals, pilot plants should be isolated from public areas, other laboratories, and office spaces by distance, special fire protection, and/or explosion-resistant construction (Jones, 1966). Frequently, special access restrictions must be imposed when inherent hazards are associated with the specialized materials and equipment being used.

6.1.2 Work Activities

The activities performed include mixing, blending, heating, cooling, distilling, filtering, absorbing, crystalizing, evaporating, grinding, size separating, and chemical reacting as a part of production or purification of a product. Some of the procedures and materials require the special ventilation capabilities that will be discussed in Section 6.3. A pilot plant usually requires a more or less permanent crew with special training that includes instruction in safety and health protection. Protective clothing and respirators must be available for all personnel working in the pilot plant.

6.1.3 Equipment and Materials Used

Analytical instruments and a full range of sensors and automatic process controllers will usually be present, in addition to large items of process equipment. Extremely hazardous materials that are sometimes used in pilot plants include chemicals of high toxicity, volatile liquids, combustible dusts, and highly explosive materials. Operations often involve the use of high voltages, very high radiofrequency generators, and other electrical equipment with a high potential for fire, explosion, and electrocution. High-pressure steam, air, and special gases are employed frequently, as are open flames, furnaces, and similar intense heat generators that make the heat load in pilot plants a special ventilation concern.

6.1.4 Special Requirements

Training of all personnel in the processes to be conducted in a pilot plant, including an understanding of first aid, emergency operations, ordinary and hazardous waste disposal, and the use of emergency equipment, should be mandatory before operations begin.

6.2 PILOT PLANT LAYOUT

Because pilot plant operations involve a large variety of equipment and operations arranged in a constantly changing pattern, it is not possible to illustrate a comprehensive layout. All of the items addressed in Sections 1.2 and 2.2 should be evaluated for their applicability, and safety and industrial hygiene personnel should be consulted for further assistance when unusual requirements are encountered. Additional sources of information include OSHA and NIOSH, State Departments of Occupational Health and Industrial Hygiene, and pertinent reference materials. Utilities provided in pilot plants, usually from a number of well distributed locations, include high-amperage single- and three-phase electrical current of 120, 240, and 440 V, compressed air to 100 psig, vacuum of $\frac{1}{2}$ atm or lower, steam at least up to 15 psig, hot and cold water, and multiple floor drains leading to waste.

6.3 HEATING, VENTILATING, AND AIR CONDITIONING

6.3.1 Introduction

Pilot plant ventilation is needed to provide an environment that is within acceptable comfort limits and to provide a reasonable capacity for diluting contaminants released into the work environment. Air change rates for chemical processing should be 5 cfm per square foot of floor area. For petroleum processes, 3 cfm per square foot is adequate. Ventilation must be provided in a manner that will not contaminate other areas of the building either by infiltration through low-pressure pathways or by contamination of air intakes with pilot plant exhaust air.

6.3.2 Additional Requirements

All of the items described in Sections 1.3 and 2.3 should be examined and all that apply to pilot plants should be implemented. Special provisions for pilot plant local exhaust systems follow.

6.3.2.1 Local Exhaust Systems

Pilot plants should be provided with fixed general laboratory ventilation systems designed to supply air at ceiling level and exhaust it from floor level. In addition, multiple local exhaust outlets should be provided from a perimeter system of main ducts by the use of quick-disconnect types of fittings and flap-type dampers that automatically close the connections when not in use. The use of such a local exhaust system is capable of serving the entire pilot plant area at the same time that it limits the amount of ventilation air exhausted, thereby conserving energy. Care in the design of these systems is necessary because a static pressure of at least 2.5 in. w.g. at each exhaust point will be required to assure adequate airflow capacity.

6.4 LOSS PREVENTION, INDUSTRIAL HYGIENE, AND PERSONAL SAFETY

All the items described in Sections 1.4 and 2.4 should be evaluated for relevance and all that apply to pilot plants should be implemented.

6.4.1 Spill Containment

Due to the larger size of operations in this laboratory, larger quantities of chemicals and other materials may be necessary. Storage containers and locations for same should be carefully selected to provide hazard separation and isolation. Spill dikes may be necessary and large quantities of spill control materials may need to be stored. Storage of these materials should be outside of the laboratory yet not isolated from it in time of need.

6.5 SPECIAL REQUIREMENTS

Unusually high ceilings combined with the use of bulky processing equipment, such as large tanks, in a constantly changing pattern make provision of uniform lighting of good quality and adequate intensity difficult. Insofar as possible, ceiling lighting should be provided by many closely spaced fixtures to avoid the heavy shadows cast by large equipment when widely spaced, high-intensity light sources are used.

Provisions should be made for the installation of an air supplied respiratory protection system. It should be a separated and protected compressed air system supplied by an oilless compressor.

7

Physics Laboratory

7.1 DESCRIPTION

7.1.1 Introduction

Research carried out in a physics laboratory may include the use of electricity (high current, voltage, and frequency levels), many chemicals (solid, liquid, and gaseous), intense light sources (lasers greater than 3 mW output power), magnetics, cryogenics, and a variety of high-energy systems including high-temperature steam and compressed air. Research may be carried out with radioactive materials to produce ionizing radiations but consideration of their use is not included in this chapter. Chapter 11 contains a description of laboratories utilizing more than trivial amounts of radioactive materials.

7.1.2 Work Activities

The basic procedures carried out in physics laboratories include experimental development of mechanical, electrical, hydraulic, and pneumatic systems and examinations of the properties of matter. Operations involve equipment setups and physical measurements for experiments that involve observation, data collection, and analysis.

7.1.3 Equipment and Materials Used

Equipment used in physics laboratories is varied and depends on the nature of the work. A partial list of physics laboratory experiments and equipment includes

Shock tube studies (air compressors, pressure-relief diaphrams, pressure and gas-flow measuring instruments).

Lasers (electrical circuits, cryogenic liquids).

Spectroscopy (carbon arc source generators, magnets, photography facilities).

Cryogenics (refrigeration equipment, liquified gases, low-temperature measuring instruments).

Electromagnetics (high electrical current services, cryogenic liquids, ion source generators).

High-frequency noise and electricity research (high-current and high-voltage electrical services).

Energy storage systems (rotary machines, heat exchangers, electrical condensers, temperature measuring instruments).

Ionizing radiation systems, including X-rays, high-current and high-intensity electrical services, and ionizing radiation measuring instruments.

High vacuum systems.

7.2 LABORATORY LAYOUT

A physics laboratory should be laid out to provide easy access to all areas within, and associated with, the laboratory by wheeled trucks, dollies, cranes, and other mechanical handling equipment. Services, such as electrical, should have easy access for safe modification of experimental set-ups. Because a physics laboratory can be used for so many different types of research, equipment configurations are likely to be different from experiment to experiment. For this reason, all services should be flexible enough to serve all parts of the laboratory conveniently. A typical layout in shown in Figure 7-1.

All the items reviewed in Sections 1.2 and 2.2 that are applicable to physics laboratories should be implemented. An additional recommendation is noted below.

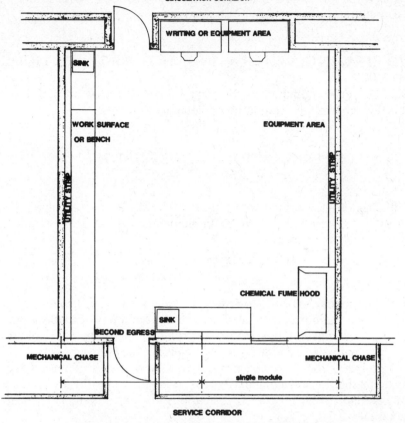

FIGURE 7-1. Physics laboratory: Sample layout.

7.2.1 Egress Routes for Physics Laboratories

In the typical physics laboratory layout shown in Figure 7-1 the two required separate egress routes are shown. The doors should be adequate in size and the route width sufficient to accommodate medical stretchers and other emergency equipment. They should also be adequate for mechanical handling equipment used for transporting heavy and bulky items.

7.2.2 Furniture Location

In laboratories utilizing unshielded laser beams, desk chairs and seated work surfaces (30–32 in.) are not recommended. Laser beams are often directed and optically transfered at seated eye level. Desks and seated computer stations should be provided in separate rooms.

7.3 HEATING, VENTILATING, AND AIR CONDITIONING

All the items contained in Sections 1.3 and 2.3 that are applicable to physics laboratories should be implemented.

7.4 LOSS PREVENTION, INDUSTRIAL HYGIENE, AND PERSONAL SAFETY

All the items contained in Sections 1.4 and 2.4 that are applicable to physics laboratories should be implemented. Additional recommendations are noted below.

7.4.1 Emergency Eyewash Stations

When chemicals hazardous to face and eyes, such as strong acids, alkalies, and other corrosive materials, are used in a physics laboratory, one or more eyewash stations must be provided according to recommendations contained in Section 2.4.1.5. Because of the large amounts of electrical equipment contained in most physics laboratories, plumbed eyewash systems should be located in a hall or some other closeby area outside the physics laboratory proper. Specifications for such a system are in Appendix VIII.

7.4.2 Fire Detection, Alarm, and Suppression Systems

Fire suppression and detection systems should be carefully planned from the earliest stages of laboratory design because even under the best of conditions, installation costs are very high for this type of laboratory.

7.4.2.1 Fire and Smoke Detection and Alarm Systems

Physics laboratories need fire and smoke detectors that respond to products of combustion (e.g., photoelectric and ionization detectors) or thermal effects (either rate of temperature rise or a final fixed temperature). Any of these detectors are adequate to sense a fire in a physics laboratory except when the laboratory uses substantial amounts of combustible gases or combustible volatile liquids. When such is the case, ionization detectors should not be used. Fixed temperature, or rate of temperature rise or photoelectric detectors are acceptable. When a physics laboratory is in a part of a building without windows to the outside and when light research, such as with lasers, is not being carried out, flame-sensing detectors are also appropriate. All detectors must be UL or FM approved and be tied into a UL- or FM-approved general building alarm system. (Section 1.4.4 contains additional information about fire detection systems.)

7.4.2.2 Fire Suppression Systems

The basic methods of fire suppression in a physics laboratory should be fixed, automatic systems combined with hand-held, easily portable extinguishers. Although sprinklers are considered to be the best fire control device for most laboratories in a research laboratory building, physics laboratories often contain special electrical hazards that should be protected with fire suppression systems other than sprinklers. These include CO_2, Halon 1301, and total-flooding dry chemical systems. An exception to this rule can be made when the laboratory is electrically deenergized upon sprinkler activation. Whenever water-sensitive equipment is not used and whenever high-voltage electronics or high-current electrical equipment, such as superconducting magnets, are not present, physics laboratories can be protected from fire spread with automatic sprinkler systems.

7.4.2.2.1 Fixed Automatic Extinguishers

Fixed automatic fire extinguisher systems used in physics laboratories should be consistent with the operations anticipated for that laboratory as explained in Section 7.4.2.2. If operations prohibit the use of a water sprinkler system, fixed automatic Halon or similar extinguisher types should be used. Of the various choices, Halon 1301 offers the following advantages: it is not life threatening because airborne concentrations are usually between 5 and 7% and it is a clean extinguishing agent that leaves no residue. Such systems must be designed, constructed, and installed in accordance with NFPA standards (NFPA 12A, 1985). Carbon dioxide systems should be avoided whenever the enclosure to be protected will be occupied. Dry chemical systems leave a powdery residue that may harm equipment. Although a powder residue is not the primary concern when selecting a fire suppression system, when there are alternatives of equal efficacy, this becomes an important criterion in the selection process.

7.4.2.2.2 Portable Extinguishers

Hand-portable fire extinguishers of adequate size that contain appropriate extinguishing agents for anticipated fires should be located in the laboratory. Appropriate hand-portable units for a physics laboratory are 15-lb CO_2 extinguishers, 2A-40 BC Halon extinguishers with Freon 1211, and 2A-40 BC multipurpose dry chemical extinguishers. Places for these extinguishers should be provided within the laboratory where they can be picked up while personnel are making an exit from the laboratory. Sizes of Halon and multipurpose extinguishers should always be in the range of 2A to 4A and 40 BC to 60BC. Portable extinguishers should also meet all the requirements of NFPA 10. (NFPA 10, 1985)

7.4.2.2.3 Special Systems
Wherever a potential for an electrical fire exists within equipment cabinets that can be protected without involving the entire building system, this type of protection should be provided with Halon systems or CO_2 total flooding systems. Such systems may be wired into the building alarm and annunciation system. All of the ventilation shutdown requirements for these special systems should be met, as outlined in NFPA 12A (Halon) or NFPA 12 (CO_2) (NFPA 12, 12A, 1985).

7.4.3 Special Equipment Requirements

Careful consideration should be given to the potential contribution of cryogenic materials, special gases, high electrical energy demands, strong laser beams, and the like, to accident, emergency, and stress situations, and design features should be considered to provide adequate barriers against damage from any of these high-intensity energy forms.

7.4.3.1 Equipment Operation with Hazardous Materials and in Hazardous Modes
Equipment that operates on materials that are hazardous (toxic or flammable) should be provided with all of the building design features, such as ventilation and emergency services, that will make it possible to control the hazards in case of unexpected events. Potential problems may be discovered by discussions between users of the laboratory and safety professionals.

7.4.4 Special Safety Requirements

When screen rooms are required to block radiofrequency energy from entering the laboratory area, entrances to these rooms should be interlocked with the electrical power source, while always permitting easy egress for personnel in the event of an emergency.

A grounding grid system should be installed in a physics laboratory to enable grounding of all pieces of equipment that are electrical in nature or that come in contact with equipment that is electrical. The grid system should be extensive enough and of sufficient size to result in only a small potential difference between the two farthest points.

Because water supply and drainage requirements may be high for water cooling of magnets or other devices, design for such systems should be evaluated early in the building planning phase. For economy of operation, a closed loop condenser water cooling system should be considered for all such facilities.

7.5 SPECIAL REQUIREMENTS

None.

8

Clean Room Laboratory

8.1 DESCRIPTION

8.1.1 Introduction

A clean room laboratory is a specially constructed and tightly enclosed work space with modulating control systems to maintain design standards for low airborne particulate matter, constant temperature, humidity, and air pressure, and well-defined airflow patterns. Clean rooms are classified by particle count per cubic foot of air. All particles 0.5 μm and larger are counted. It is important that the level of cleanliness that is to be maintained be determined from the beginning of the design process. Three levels of cleanliness are recognized in clean room practice: rooms containing no more than (1) 100,000, (2) 10,000 and (3) 100 particles per cubic foot. The cleanliness designations are derived from Federal Standard 209b (GSA, 1979). Although not included in the Federal Standard, clean rooms designed for particle counts of no more than 10 per cubic foot are becoming common in the microchip industry, and clean rooms designed for no more than one particle per cubic foot of air are in the initiation stage. This chapter will deal with Class 100 clean room laboratories that permit no more than 100 particles per cubic foot of laboratory air.

8.1.2 Work Activities

The activities performed in a clean room laboratory are characterized by a need for extreme cleanliness rather than by the nature of the activities.

Experimentation with, and development of, electronic microchips, miniature gyroscopes and switches for guidance systems, and photographic films and film processing techniques are samples of the kinds of work that require ultraclean laboratories for successful operations. Chemical treatment, precision machining, solvent cleaning of parts and mechanisms, and use of a wide spectrum of precision measuring devices are activities characteristic of clean room laboratories as a class. A clean room laboratory requires all of the special ventilation and air filtration facilities that are described in Section 8.3. The clean room laboratory may also be used for storage of materials that require a high degree of cleanliness as well as temperature and humidity control in accordance with a variety of experimental operating procedures. The laboratory will have access restrictions, and donning of special lint-free clothing will be required for entrance into the clean room laboratory as an aid to maintaining the required low dust level.

8.1.3 Special Requirements

Clean room laboratories may contain multiple rooms, each with different requirements for contamination control. Each room within the clean room laboratory should be maintained at a static pressure higher than atmospheric and higher than that in adjacent indoor spaces to prevent air infiltration from less well controlled areas. Differential pressures should be maintained between adjacent rooms of the multicompartmented clean room laboratory to assure airflow outward from the cleanest spaces to those maintained at a lesser standard of air dustiness. See all of Section 8.3 on this subject.

8.1.3.1 Structural Materials

Clean room laboratories should be constructed of smooth, monolithic, easily cleanable materials that are resistant to chipping and flaking. The interior surfaces should have a minimum of seams and be devoid of crevices and mouldings. Walls should be faced with plastic sheeting or covered with baked enamel, epoxy, or polyester coatings with a minimum of projections. Ceilings should be of metal or plastic-faced panels or covered with plastic-finished acoustical ceiling tiles. Floors should be covered with sheet vinyl or painted with an epoxy application that will form a monolithic surface with gently rounded cove base.

8.1.3.2 Personal Cleanliness

Personnel practices are very important in clean room laboratory operations. To ensure cleanliness, personnel should be provided with lint-free smocks, gloves, shoe covers, and head covers, and have available wash areas with soap or lotions containing lanolin to tighten the skin and thereby reduce sloughing of skin fragments. All equipment and materials should be thoroughly cleaned before being brought into a clean room laboratory.

8.1.3.3 Auxilliary Fans

The high ventilation rates required to maintain air cleanliness may dictate a need for auxilliary fan rooms, and special attention must be given to their design, cleanliness, and maintenance requirements. See ASHRAE, 1982 for additional information.

8.2 LABORATORY LAYOUT

8.2.1 Introduction

The layout of a clean room laboratory may be a single room, with a dust-free air-pressurized vestibule, or it may be a laboratory suite of several separate rooms interconnected by an internal, pressurized corridor. For emergency evacuation, the internal corridor should connect to a building egress through pressurized or nonpressurized vestibules. A clean-air-pressurized robing room is an essential adjunct of a clean room laboratory. A pressurized vestibule may be used for this purpose if there is adequate space for storage of clothing and installation of a handwashing sink, a shoe cleaner, and/or a sticky shoe mat. Otherwise, a separate robing room must be provided for in the plan. When a robing room separate from a pressurized vestibule is used, it should be at a higher air pressure than the access corridor, but be at a lower air pressure than the vestibule and clean laboratory rooms. Access to the clean laboratory room will be through an air-blast chamber placed between the robing room and the laboratory rooms. Its purpose is to blow lint, skin fragments, hair, and so forth, from personnel prior to their entry into the cleanest areas of the suite. The minimum dimension of a clean laboratory room work space should be 7 ft 6 in. A typical clean room is shown in Figure 8-1. The items contained in Sections 1.2 and 2.2 apply to clean room laboratories except when modified in the following sections.

8.2.2 Ingress and Egress

8.2.2.1 Egress Routes

A minimum of two separate egress routes from each clean room laboratory unit is recommended and the exits should be as far apart as possible and lead to different fire zones as a safety measure. In addition, there should be no dead-end corridor longer than 20 ft when measured from the

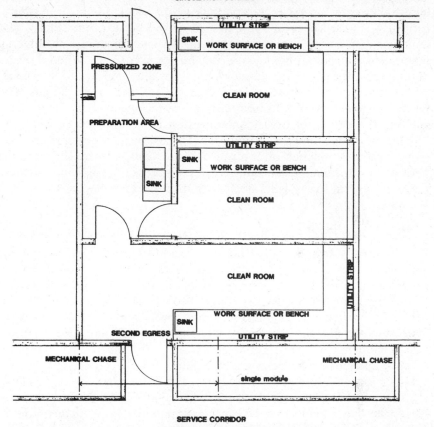

FIGURE 8-1. Clean room suite: Sample layout.

centerline of the door of the farthest room to the door that opens onto a building egress corridor. A second exit may be labeled as an emergency exit and audibly alarmed to deter use except under emergency conditions. An approved crash bar or "California firelock" hardware will allow emergency egress but limit unauthorized entry. When particulate contamination from outside the clean room laboratory unit is controlled by pressurized anterooms and robing rooms, a one-way flow of traffic from a separate entrance to a separate exit is not essential.

8.2.2.2 Traffic Flow

In situations where contamination generated or contained within the clean room must be prevented from leaving the area, a restricted one-way flow of traffic may be needed. This calls for a place where personnel may change into protective garments, or put them over their street clothes, before they enter a pressurized vestibule giving access to the clean laboratories. A similar space at the exit end of the clean room facility that includes wash basins or showers, and an autoclave for sterilizing outgoing materials and protective garments, may be needed as well. When showers are required, the one-way internal corridor should loop back to the changing room in order for personnel to gain access to their street clothes. The project engineer should discuss with industrial hygiene and safety consultants the degree of cross-contamination to be allowed between the building air supply and the clean room. There are many possible arrangements whereby a series of controlled spaces can be made to meet the various air cleanliness and interlocking air pressure gradient specifications. Biological safety laboratories are discussed in Chapter 12.

8.2.2.3 *Interlocking Doors in Ingress and Egress Pathways*

It is frequently proposed that access corridors, pressurized vestibules, and robing rooms associated with clean rooms and clean room laboratories be provided with interlocking doors so that personnel will be prevented from opening both doors simultaneously and, by this means, inadvertently introduce contamination into clean areas. This is a very dangerous arrangement because under the stress of a fire, explosion, or other emergency in a clean room laboratory, it is unlikely that an orderly opening and closing of doors will take place. People can become trapped between locked doors in a vestibule or robing room under panic-producing situations. It is acceptable to place visual and audible alarms on double-access doors to alert supervisory personnel to failures in dust-control discipline so they may take prompt corrective measures, but no barriers to emergency egress should be permitted.

8.3 HEATING, VENTILATING, AND AIR CONDITIONING

8.3.1 Introduction

The HVAC equipment that will be needed for a clean room laboratory depends on the clean room class, temperature and humidity requirements, the need for fume hoods and spot exhaust hoses, and the presence of major heat-generating equipment and activities. The items contained in Sections 1.3 and 2.3 apply to clean room laboratories and should be implemented. In addition, the following provisions should be considered for inclusion in building and laboratory plans.

8.3.2 Environmental Control

8.3.2.1 Ventilation

Good ventilation is critical in a clean room. Large volumes of dust-free air and excellent air velocity and directional control are needed to provide the required cleanliness level. The ventilation system for the clean room should be independent of the general building supply system. When toxic chemicals are to be used, local exhaust or chemical fume hoods will be needed to meet the provisions of Section 2.3. Information on ventilation of toxic gas cylinder storage cabinets has been published (Burgess, 1985).

8.3.2.2 Filtration

Filtration is required for all outside supply air (called primary air) or recirculated air (called secondary air) to a clean room laboratory. Primary air should be filtered: first, by a roughing filter capable of removing coarse particles and fibers; second, by a prefilter of 85–95% atmospheric dust efficiency; and third, by a HEPA filter, efficiency rated at 99.97% or higher. Secondary air is usually filtered only through HEPA filters. In a clean room laboratory, the supply air may be delivered to a ceiling plenum containing a continuous bank of HEPA filters to provide filtered air circulation downward to the floor where there will be return air grilles. A fraction of the air will be exhausted to the outside, the remainder will be recirculated through the HEPA filters. A similar arrangement is possible when horizontal airflow is desired by building a wall of HEPA filters at one end of the room and locating return grilles in the opposite wall. These ventilation plans are illustrated in Figure 8-2.

8.3.2.3 Room Pressure Balance

Pressure control within the clean room should be maintained by static pressure controllers that operate dampers, fan inlet vanes, or a combination of both to maintain the correct ratio of supply to return and exhaust air. To provide close control of room pressure, disturbances to airflow should be minimized by operating fume hoods and local exhaust points continuously and providing air locks between adjoining areas. The specifications for maintaining room pressure balance should be the same as those required by the General Services Administration (Fed. Std. 209b) for clean rooms processing government contracts. In certain applications simplified systems without the use of separate primary/secondary fan systems may be sufficient. See ASHRAE, 1984, for an additional reference.

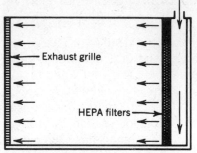

Horizontal laminar flow clean room

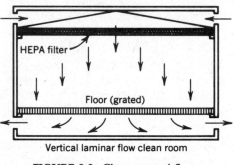

Vertical laminar flow clean room

FIGURE 8-2. Clean room airflow.

8.3.2.4 Malfunction Alarms

Alarms should be provided to indicate a malfunction of the directional airflow arrangement. Additional alarms should be provided for the filter banks to notify service personnel that the filters are becoming loaded with dust to a point where airflow delivery is being affected adversely. It should be kept in mind that it is necessary to measure total airflow rate as well as filter bank pressure drop to evaluate the condition of the filters correctly.

8.3.2.5 Humidity Control

It is important that humidity and temperature conditions be controlled. Humidity control is required for corrosion and condensation control, static electricity eliminations, and personal comfort. Temperature control provides stable conditions of operation for instruments and personnel.

ASHRAE 1982, Chapter 16 discusses this issue in more detail. In general, conditions to be met are

72°F ± 2°F temperature

45% ± 5% relative humidity.

In critical applications, tolerances of ±0.5°F and ±2% RH may be required. If a relative humidity of less than 35% is to be maintained, special precautions plus static electricity control must be taken.

8.4 LOSS, PREVENTION, INDUSTRIAL HYGIENE, AND PERSONAL SAFETY

The information contained in Sections 1.4 and 2.4 applies to clean room laboratories and should be implemented, with the following addition.

8.4.1 Portable Fire Extinguishers

In addition to the fire extinguisher recommendations contained in Section 1.4.4.2.2, space should be provided during the planning stage for hand-portable fire extinguishers to be used in the event of a filter fire. The size and type of extinguishers should be determined with the aid of a safety professional.

8.5 SPECIAL REQUIREMENTS

8.5.1 Lighting

Clean room laboratory lighting should be 100 ft-c at 30 in. above the floor to provide a reasonable illumination level for fine, precision work that will avoid eye fatigue.

9

Controlled Environment (Hot or Cold) Room

9.1 DESCRIPTION

9.1.1 Introduction

A controlled environment room is a laboratory or a laboratory adjunct in which temperature and humidity are maintained within a specified range so that laboratory activities can be conducted and laboratory products maintained under controlled conditions. A controlled environment room can be maintained to within 1°F at an elevated temperature up to 120°F, or at a reduced temperature down to 35°F. In addition, relative humidity can be controlled to within 0.5% of the full humidity span.

9.1.2 Work Activities

Although controlled environment rooms are primarily designed and used for storage of sensitive materials that require maintenance within a specified temperature and relative humidity range, they are frequently used, in addition, for conducting activities normally performed in a general chemistry laboratory or in biology, bacteriology, or cell culture laboratories.

9.1.3 Equipment and Materials Used

It is expected that few pieces of analytical equipment will be located permanently in controlled environment rooms because of the unfavorable storage environment. Specialized storage containers will usually be found there. Controlled environment rooms usually have no special access restrictions unless sensitive equipment or dangerous materials are being used therein.

9.1.4 Exclusions

Those cited in Section 3.1.4 apply to controlled environment rooms.

9.1.5 Special Requirements

Ideally, environmental rooms should be the product of a single manufacturer and completely furnished and installed by the same manufacturer to avoid division of responsibility. If possible, rooms should be prebuilt at the manufacturer's plant and pretested prior to shipment. Pretesting conditions should simulate, or even exaggerate, the environmental conditions that will be found in actual service. Tests should include a thorough check of the mechanical, electrical, and temperature-control systems. It is advisable to have an owner representative present during the manufacturer's preinstallation testing program.

9.2 LABORATORY LAYOUT

9.2.1 Introduction

A controlled environment room is a laboratory type that need not always conform to the dimensional guidelines set out in Section 1.2.2.3. The interior area of manufacturers' standard controlled environment rooms can vary from closet size (20 net square feet (nsf)), to double-laboratory module size (400 nsf), or even larger. The minimum dimension may be less than 7 ft 6 in. because a controlled environment room is not usually classified as a habitable room. If a controlled environment room larger than double laboratory size is needed, safety personnel should be consulted for assistance in the preparation of design specifications.

Temperature conditions within controlled environment rooms are usually outside the comfort zone for personnel, so work efficiency must be carefully considered during the layout of work surfaces and storage units. When work surfaces and sinks are present inside the controlled environment room, they should be located close to the door so that the zone of greatest activity will be near the exit. Because space is usually limited in this type of laboratory facility, storage units should be placed to the rear

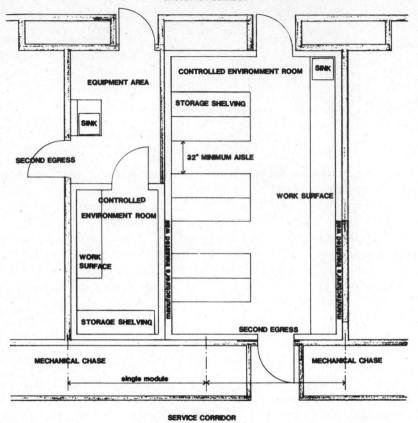

FIGURE 9-1. Controlled environment room: Sample layouts.

so that materials on and in them will be less likely to get bumped and spilled.

Examples of large and small environmental rooms are shown in Figure 9-1.

9.2.2 Egress

9.2.2.1 Doors

Two doors are not practical or required in controlled environment rooms less than 200 nsf, because of limited wall perimeter. However, controlled environment rooms of 400 nsf and over should have a second egress. For rooms between 200 and 400 nsf, a second egress should be considered on the basis of individual need.

9.2.3 Furniture Location

Work surfaces and storage shelving should be designed and located to facilitate ease of egress. Other furnishings should be kept to a minimum, so that aisles are not obstructed. In controlled environment rooms having a minimum width of 15 ft, an island-type work surface, bench, sink, or storage unit may be located inside the room, provided a minimum aisle width of 36 in. is maintained on all sides to facilitate egress. Aisles between parallel rows of work surfaces or storage units, and aisles between an interior wall and a work surface or storage unit should be not less than 36 in. Desks should not be placed in controlled environment rooms.

9.2.4 Access for Handicapped Persons

To make controlled environment rooms accessible to persons in wheelchairs, a clear floor area of at least 5×5 ft will be required for turning the wheelchair around, and a work surface 32 in. high should be provided.

9.3 HEATING, VENTILATING, AND AIR CONDITIONING

9.3.1 Introduction

HVAC requirements for a controlled environment room depend on the temperature and humidity conditions to be maintained, as well as on the activities to be performed. Outside air volume rates should be kept to a minimum, particularly when close humidity control is required. Most cooling systems are direct expansion refrigeration type with separate condensing unit and evaporator coil section.

Unless there are special requirements, such as a need for a fume hood or numerous local exhaust points, outside air exchange requirements for controlled environment rooms will be minimal. A minimum of 50 cfm of outside conditioned air is recommended when people must work inside the Controlled Environment Room regularly for prolonged periods of time. To provide fresh filtered air for the people using the room, a supply air blower with full-modulating control should be connected to a ceiling plenum located at the entrance to the work area and the supply air should be discharged directly through the evaporator or heating coil. Air should be exhausted to the building exhaust air system through an adjustable damper. Supply and exhaust air volumes will depend on room size and the activities to be conducted inside the controlled atmosphere room. When toxic chemicals will be used, spot exhaust points or chemical fume hoods will be needed, and they should conform to the recommendations in Section 2.3.4.3.

9.3.2 Temperature and Humidity

9.3.2.1 Room Temperature

For low-temperature control, the refrigeration system should contain a direct expansion unit of industrial quality, designed to operate continuously with an integral evaporator coil. Room temperature controller and other instrumentation should be designed to control coil temperature over the full temperature range on a precise demand basis. The control mode should be fully modulating with proportional action from 0 to 100% of total condensing unit capacity over the full-rated temperature range. The main controller should include a means for direct setting of the control point, input and output meters to display the proportioning action of the control unit, a temperature indicator to permit monitoring room temperature conditions with a set-point accuracy of not less than 0.5% of the full-rated temperature span, and a proportioning band of not less than 1% above the central point. The temperature-sensing unit must possess adequate inherent stability, accuracy, and sensitivity to provide the degree of temperature control required for the operations that will be conducted in the controlled environment room. A recorder with a 12-in. circular 7-day or 24-h chart should be installed in a central control panel to assist in monitoring the stability of the set conditions.

9.3.2.2 Room Humidity

When humidity control is required, clean, dry, low-pressure steam from a separate, dedicated service should be discharged to the room in response to a pneumatic proportioning control system. The controller should be fully calibrated and include an electronic sensing unit, an integrated recorder with a 12-in. recording chart, and a set-point accuracy of not less than 0.5% for the full humidity span between.

9.3.3 Emergency Alarm and Control System

A safety control and alarm system, provided by the manufacturer, should be mounted on an outside wall of the cold or warm room adjacent to the entrance door. It should consist of an independent electrical low- and high-temperature control system that will take over operation in the event of a main control failure and contain an alarm buzzer to give audible warning in the event of temperature deviations. The safety control and alarm should be equipped with the following components: a main on/off switch for the entire system, a silencing switch for the buzzer, and a reset switch to return the system to normal operation. Terminals should be provided to connect the alarm system into a remote central location.

9.4 LOSS PREVENTION

Information contained in Sections 1.4 and 2.4 applies to all controlled environment rooms and all relevant items should be implemented. In addition, when an experiment could give rise to a hazardous situation, either by depleting oxygen or by releasing toxic contaminants, provisions should be made during the building design phase to install facilities for supplied air or self-contained breathing apparatus, plus one or more atmospheric monitors to identify the hazardous situation and provide an appropriate alarm.

9.5 SPECIAL REQUIREMENTS

9.5.1 Materials of Construction

Controlled environment rooms may contain a rapidly degrading environment for some of the services and equipment installed in it. For example, hot rooms can reduce the normal expected life of electrical wiring through deterioration of the insulation. The National Electrical Code should be consulted to determine whether an over-design in wire size and insulation type will be required due to the specific temperatures and current loads that will be encountered.

Frequently, cold rooms become wet with moisture that condenses from building air that enters the room when people go in and out. The moisture affects materials of construction such as steel shelves, electrical conduits and fixtures, and moisture-absorbing materials. The use of a vapor-tight electrical system can help prevent early deterioration and possible shock hazards. A careful selection of materials should result in nonrusting shelves and other structural components of the cold room.

9.5.2 Lighting

Interior lighting should be high-output fluorescent designed to provide 70 ft-c evenly distributed when measured at 40 in. above the floor. Ballasts should be mounted externally and fluorescent lamps should have moisture-proof covered socket ends. Lens panels should be the diffuser type made from acrylic.

10

High-Pressure Laboratory

10.1 DESCRIPTION

10.1.1 Introduction

A high-pressure laboratory is designed, constructed, and operated to permit safe experiments at gas pressures over 250 psig and liquid pressures over 5000 psig.

Because of the high-energy potential of high-pressure fluid systems, special consideration must be given to the location of the laboratory within the building structure and to its materials of construction. For example, a 10-ft³ volume of dry nitrogen at 6000 psig has the energy equivalent of approximately 300 lb of TNT. Designing a laboratory to handle a potential explosion of this magnitude requires great care. Ideally, the laboratory should be a free-standing barricaded building, but when it is located inside a laboratory building, the ultimate in control of the qualities and quantities of materials that will be used in its construction and management of laboratory procedures will be required.

10.1.2 Work Activities

The investigative procedures used in a high-pressure laboratory include those used in a general chemistry laboratory except that they will be conducted on gases and liquids at higher pressures. Pressure testing of vessels, high-pressure reactions, and some handling of very high temperature, as well as cryogenic, liquids may take place in the high-pressure laboratory. Small quantities of high explosives may also be used.

10.1.3 Equipment and Materials Used

High-pressure oil pumps, high-pressure gas compressors, high-pressure piping and valves, accumulators, barricades, and compressed gas cylinders are used in high-pressure laboratories. There is likely to be a need for some chemical storage. The location of each of these items should be considered carefully from the safety standpoint. Because much of the equipment is necessarily heavy, floor loading requirements must be evaluated early in the building planning process. Access restrictions will apply to this laboratory during many operating conditions.

10.1.4 Exclusions

High-pressure energy systems that exceed the design capabilities of the facilities must be carried out in other locations. Prior to construction, the designer/architect/engineer must determine the maximum size of experiments and pressure conditions that will be allowed in the facility, and management controls must be instituted to make certain these limits are never exceeded.

10.2 LABORATORY LAYOUT

10.2.1 Introduction

All provisions of Section 2.2 apply to high-pressure laboratories and should be implemented. In addition, wall, floor, and ceiling construction should be designed to contain the explosive violence of maximum energy release. One method of construction is to contain the released energy; an alternative method is to channel the release of energy into safe pathways designed to protect persons and property outside the laboratory.

Full containment of an explosion requires the construction of high-pressure areas, or "cells," along outside walls. The walls and ceiling of cells are constructed of reinforced 18-in. thick concrete, or of steel, or of combinations of materials of adequate strength to contain the maximum anticipated explosion. All exterior approaches to the cells must be secured. The cell is sealed with a door assembly of required strength and fire rating. Personnel conducting experiments in which explosions are anticipated are restricted from the active cell during the actual experiment.

Venting panels may be installed in a wall or, if the high-pressure laboratory is in a single-story building, in the roof to direct the safe release of explosive energy. The panels are designed to open outward by the force of a sudden pressure rise before structural damage occurs. The free area required and release pressure are specified in NFPA 68 (NFPA 68, 1985) for laboratories according to the amounts and nature of the explosive or volatile materials used. The area outside the blowout or pressure relief panels should be protected as described in the following paragraph.

Locations for blowout panels and cell walls should be selected in locations that prevent passersby from close approach. Blowout panels should be securely tethered to immovable elements of the building structure with loose ropes of steel cable to prevent their release as free projectiles following a forceful explosion. In addition, heavy mesh screening or a similar barrier should be installed to contain the debris that may be ejected through the opening. Earth berms of sufficient height and breadth are the best means for containment, but if there is not enough area outside the high-pressure laboratory for a protective berm, a solid blank wall with the strength to withstand the blast should be constructed. As a minimum, a high, heavy-duty chain link fence should be erected far enough away to prevent any dangerous approach to the panels or cells.

If the high-pressure laboratory is within a multistory building, it should be located on the ground floor. Windows or other building openings above blowout panels should be protected from possible flame spread and explosion debris by a solid canopy above the blowout panel of sufficient strength and fire rating to withstand the hazards.

All laboratories and storage rooms involving a significant explosion hazard must comply with NFPA, local building codes, and fire department regulations.

The construction of a fully equipped high-pressure laboratory can be avoided by substituting a specially designed, heavy-walled vessel that is large enough to house the experiment plus auxilliary equipment and strong enough to contain the energy resulting from any accident. For specific details on construction of heavy-walled pressure vessels, the ASME unfired Pressure Vessel Code, Section VIII (ASME, 1983) should be consulted.

10.3 HEATING, VENTILATING, AND AIR CONDITIONING

10.3.1 Special Requirements

A high-pressure laboratory presents a number of unique problems for the designer. High-pressure laboratories are seldom air conditioned, but when they are, the ducts and fans may have to be specially constructed and mounted to withstand explosions. Also, special venting arrangements must be provided to relieve excessive pressure safely under unexpected circumstances.

When a high-pressure laboratory functions at ordinary pressures, all the provisions of Sections 1.3 and 2.3 that are applicable to its new use should be implemented, plus those that apply to whatever type of laboratory it most closely resembles.

Ventilation rates of 30–45 air changes per hour, characteristic of pressure cells, are considered more than adequate for most alternative operations, but the use of more than trace amounts of highly toxic chemicals will require reevaluation of ventilation rates.

10.4 LOSS PREVENTION, INDUSTRIAL HYGIENE, AND PERSONAL SAFETY

All the items contained in Sections 1.4 and 2.4 apply to high-pressure laboratories and should be implemented. In addition, the following recommendations should be given careful consideration.

10.4.1 Portable Fire Extinguishers

Space should be provided for hand-portable fire extinguishers of larger capacity than those used in other laboratory types. They should be sized and typed in consultation with a safety specialist to make certain they will provide adequate protection for the planned operations.

10.4.2 Compressed Gas Cylinder Racks

Racks similar to those described in Section 2.4.7 should be located so that an explosive incident within the high-pressure laboratory will not spray the storage area with schrapnel that may puncture other tanks and cause the explosive release of additional toxic or flammable materials.

10.5 SPECIAL REQUIREMENTS

None.

11

Radiation Laboratory

11.1 DESCRIPTION

11.1.1 Introduction

A radiation laboratory is designed and constructed to provide a safe and efficient work place for a wide variety of activities associated with materials that may produce ionizing radiation in either the electromagnetic spectrum or in gaseous or particulate form (i.e., radioisotopes). General chemistry activities may also be performed in this laboratory. It is preferable that this type of laboratory be located in a building containing similar laboratory units but not necessarily all conducting work with radiation producing materials. Not all of the design considerations presented here may be needed in every radiation laboratory. Close communication between users and designers will be necessary to determine when certain items can be omitted with safety.

11.1.2 Work Activities

The basic experimental procedures used in this laboratory will be similar to those conducted in general chemistry and analytical laboratories but special provisions must be made for the additional safety procedures that will be required when handling radioactive materials or radiation-producing equipment. Although this chapter will describe a facility similar to a general chemistry laboratory, the safety requirements that will be outlined can be applied to other laboratory types, such as chemical engineering, physics, or high-toxicity laboratories, in which radioactive materials will be used. Any laboratory that uses radioactive materials with a total energy greater than 1 μCi should be considered a radiation laboratory.

11.1.3 Equipment and Materials Used

The equipment used in a radiation laboratory will depend on the specific nature of the work but, in general, it will be similar to that found in a general or analytical chemistry laboratory. X-ray producing equipment may be included.

11.1.4 Exclusions

This section does not cover the use of ionizing radiation in connection with animal studies or biological agents. It does not cover activities that produce non-ionizing radiation.

11.1.5 Special Requirements

A radiation laboratory should have special access restrictions. In addition to permits and licenses, there are stringent EPA, OSHA, and NRC requirements pertaining to acquisition, storage, use, release, and disposal of radioactive materials that must be followed. In some cases, provisions must be made for change rooms and showers. Health and safety professionals, including health physics specialists, should be consulted when the amounts of radioactive materials or energy levels exceed minimal levels.

Energy emitting devices should be directed away from laboratory entrances and primary egress aisles.

11.2 LABORATORY LAYOUT

11.2.1 Introduction

Many laboratory layouts are acceptable. Final choice will depend largely on the type of work to be performed and available space. A radiation laboratory layout similar to a general chemistry laboratory, as well as a number of specialized work utilization layouts, are described in *Safe Handling of Chemical Carcinogens, Mutagens, Teratogens, and Highly Toxic Substances* (Walters, 1980). Whatever laboratory design is selected, all the provisions of common elements of laboratory design contained in Sections 1.2 and 2.2 apply and should be followed. In addition, the items that follow should be reviewed for their applicability to the particular laboratory work being planned. Because the need for some of the special items will depend on the nature of the work and the quantity of radioactive materials that will be used, health and safety professionals should be consulted for assistance.

11.2.2 Change, Decontamination, and Shower Rooms

Adequate facilities for laboratory workers to change and shower must be provided when procedural requirements call for the use of protective clothing. For the use of some substances, OSHA and NRC requirements specify that a shower be available as a part of the radiation laboratory, whereas the use of other radioactive chemicals necessitates only that there be a shower available in the building.

11.2.3 Work Surfaces

Work surfaces should be smooth, easily cleanable, and constructed from impervious materials such as stainless steel with cove joints. Strippable epoxy-type paint can be used as a substitute. The use of an impervious and disposable bench covering during work should be considered as an aid in the cleanup of radioactive spills.

11.2.4 Floors and Walls

In areas using liquid or particulate radioisotopes, floor coverings should be of monolithic materials, such as seamless vinyl or epoxy. Cracks in floors, walls, and ceilings should be sealed with epoxy or a similar non-hardening material. All penetrations of walls, floors, and ceilings by utility conduits should be similarly sealed with epoxy. Lighting should be provided by sealed vapor- and waterproof units, and lighting fixtures should be flush with the ceiling to eliminate dust collection.

11.2.5 Handwashing Facilities

Handwashing facilities should be located within the radiation laboratory and in change and shower rooms.

11.2.6 Access Restrictions

Entrances to the laboratory should be posted with permanent signs indicating restricted access due to the use of a specific class of radioactive materials or radiation producing equipment (e.g., an X-ray unit). The use of special key access should be considered. Because of access restrictions and isolation of the laboratory, it is recommended that there be a viewing window in the doors. Such windows should not be of such size and construction that they compromise design fire ratings.

11.3 HEATING, VENTILATING, AND AIR CONDITIONING

11.3.1 Introduction

The items contained in Section 2.3 apply to radiation laboratories and all that are relevant should be implemented. Additional considerations are given below. Industrial hygiene or radiation safety professionals should be consulted for assistance when large amounts of radioactive materials or high ionizing radiation-producing devices will be used.

11.3.2 Fume Hood

The NRC recommends an exhaust volume of 150 cfm/ft^2 of maximum open-face area for fume hoods handling radioactive materials. Where not required by regulation we recommend 100 cfm/ft^2 of maximum open-face area. An isotope fume hood with cleanable surfaces (Appendix Vd) should be used.

11.3.3 Glove Box

Work with volatile and powdery radioactive materials calls for the use of a completely enclosed, ventilated work space, called a glove box or gloved box, rather than an isotope fume hood. The glove box should meet the specifications of the American Conference of Governmental Industrial Hygienists (ACGIH, 1984) and those outlined in Appendix Ve. Local ventilation at the entrance port may be required. The glove box is a closed system and should be kept under a negative pressure of 0.25 in. w.g. static pressure relative to the laboratory. The laboratory itself may be under a negative pressure of 0.05 in. w.g. relative to the atmosphere to prevent exfiltration of potentially contaminated air, making the box interior 0.3 in. w.g. lower than atmospheric pressure.

11.3.4 Spot Exhaust for Instruments

Analytical instruments that are used to weigh or manipulate radioactive chemicals should be equipped with spot exhaust ventilation at potential points of contaminant release, or be completely enclosed in a ventilated enclosure such as a glove box or an isotope fume hood. Specific design requirements will vary with each type of equipment and chemical used. Consultation with industrial hygienists and health physicists is highly recommended when amounts of radioactive isotopes or energy levels are greater than minimal.

11.3.5 Storage Facilities

Cabinets, refrigerators, and all other equipment items and areas used for storage of radioactive materials should be provided with local or general exhaust ventilation sufficient to maintain a directional airflow and prevent buildup of radioactive contaminants within the storage space. Laboratory users need to be consulted to determine whether any radioactive materials that emit radioactive gases will be stored in the facility. Sections 1.4.7 and 2.4.6 contain design information for chemical storage facilities.

11.3.6 Filtration of Exhaust Air

Sometimes air exhausted from isotope fume hoods, glove boxes, and spot exhaust systems must be filtered before release to the environment to avoid atmospheric pollution and to conform with NRC regulations. Industrial hygienists and health physicists should be consulted for an evaluation of this need. The first stage of filtration should be a HEPA filter with not less than 99.97% efficiency for 0.3-μm particles. A second stage containing an adsorbent such as activated charcoal may be required when the radioactive effluents are gases or vapors. The components of the air cleaning train and the size of elements will depend on the materials to be used and the total quantity of air to be exhausted. All air cleaning elements should be capable of being changed without exposure of maintenance personnel (e.g., bag-in, bag-out procedures).

11.3.7 Exhaust Stream Monitors

The Nuclear Regulatory Commission (NRC) requires continuous air monitoring of exhaust air streams for some radioisotopes when the concentrations exceed predetermined levels. Access for sampling locations must be provided in the initial design when it is anticipated that effluent air monitoring may be needed.

11.4 LOSS PREVENTION, INDUSTRIAL HYGIENE, AND PERSONAL SAFETY

The recommendations contained in Sections 1.4 and 2.4 apply to radioactive laboratories and should be followed.

11.4.1 Radioactive Waste

When special procedures for handling radiation waste are required, provisions for separate radiation waste storage areas and easy access to shipping areas must be made in the building plans. All provisions of Section 1.4.6 apply to radioactive waste storage facilities plus the use of radioactive shielding when required.

11.5 SPECIAL CONSIDERATIONS

11.5.1 Shielding

Some radiation producing equipment and radioactive materials require the use of special shielding. Generally, cobalt and cesium irradiators should be placed in separate, small rooms that are lead lined or constructed with dense concrete walls, floors, and ceilings. These facilities add many tons of extra weight to the structure and require special consideration during structural design.

12

Biosafety Laboratory

12.1 DESCRIPTION

12.1.1 Introduction

Accidental contact with infectious and toxic biological agents is an occupational hazard for persons who work in laboratories handling oncogenic viruses, infectious agents, and similar harmful biological substances. Often, research activities with these substances call for additions of radioactive tracers and chemicals that are known to be mutagenic, teratogenic, or carcinogenic, thereby substantially increasing the hazardous nature of the work. As of 1978, the registry of laboratory-acquired infections listed in excess of 4000 cases and it has been estimated that this number represents only a fraction of the total, because reporting of these cases is not compulsory.

The Center for Disease Control, an agency of the U.S. Public Health Service, has classified laboratories handling hazardous biological agents into the following four biosafety categories:

Biosafety Level 1. Suitable for experiments involving agents of no known, or of minimal, potential hazard to laboratory personnel and the environment. The laboratory need not be separated from the general traffic patterns of the building and work is generally conducted on open bench tops. Special containment equipment is not required or generally used.

Most general purpose laboratories, such as a general chemistry laboratory (Chapter 3), are suitable for work at biosafety level 1.

Biosafety Level 2. Suitable for experiments involving agents of moderate potential hazard to personnel and the environment. Access to the laboratory should be limited when experiments are being conducted. Procedures involving large volumes, or high concentrations, of agents, or when aerosols are likely to be created, are conducted in biological safety cabinets or other physical containment equipment.

Biosafety Level 3. Suitable for experiments involving agents of high potential risk to personnel and the environment. Access to the laboratory is controlled at all times. The laboratory must be equipped with special engineering and design features plus physical containment equipment and devices. All procedures involving the manipulation of infectious material must be conducted within biological safety cabinets or similar physical containment devices, or by personnel wearing appropriate personal protective clothing and devices.

Biosafety Level 4. Required for experiments involving agents that are extremely hazardous to laboratory personnel or that are exotic to the United States. Laboratory staff must understand the primary and secondary containment function of standard and special practices, the containment equipment, and the laboratory design characteristics. Access to the laboratory is strictly controlled. The facility is either in a separate building or in a controlled area within a building that is completely isolated from all other areas of the building.

The publication *Biosafety in Microbiological and Biomedical Laboratories,* issued by the Centers for Disease Control, (CDC, 1984), should be consulted to determine the biosafety laboratory level required for specific etiological (infectious) agents, oncogenic viruses, and recombinant DNA strains. This chapter deals only with laboratory design for safe handling of microbiological agents requiring a biosafety level 3 facility. A biosafety level 1 laboratory requires no special provisions and a biosafety level 4 laboratory requires facilities that should be designed and constructed only by specialists in this technology. The special construction and facilities of a biosafety level 2 laboratory are all covered in the biosafety level 3 laboratory description.

Level 3 biological safety laboratories are designed as secondary containment facilities by the creation of physical enclosures and negative air pressures and include within them primary containment facilities in the form of biological safety cabinets in which hazardous operations are conducted.

12.1.2 Work Activities

The activities performed in a biosafety level 3 laboratory include work with low- to moderate-risk biological agents. Often, microgram quantities of radionuclides, carcinogens, teratogens, and mutagens are employed as adjuncts to work with biological agents but in no case is it intended that the biosafety level 3 laboratory will be used for work with large quantities of gaseous and liquid chemicals that are hazardous by virtue of their toxicity, radioactivity, or flammability.

12.1.3 Equipment and Materials Used

The unique equipment item in biosafety laboratories is the biosafety cabinet, designed to provide personnel, product, and environmental protection when working with hazardous biological agents. The most widely used biological safety cabinet is officially designated "Class II (Laminar Flow) Biohazard Cabinetry" and its construction, performance, and testing are covered by National Sanitation Foundation Standard No. 49 (NSF, 1983). Class I cabinets provide personnel and environmental protection but no work protection, and for that reason are seldom used for work with biological agents that must be protected from contamination. Class III cabinets are total enclosure, negative pressure glove boxes equipped with at least one double-door lock containing a facility for decontaminating items withdrawn from the cabinet. Class III cabinets provide the highest level of personnel, work, and environmental protection, and find use in biosafety level 4 laboratories.

Additional equipment items found in biological safety laboratories include centrifuges, high-speed blenders, sonicators, and lyophilizers. Each of these equipment items has an ability to generate large numbers of respirable aerosolized particles that represent a potential infective dose when working with hazardous biological agents. Incubators, sterilizers, and refrigerators are also commonly found in biohazard laboratories. Specialized analytical devices that are likely to be present include gas and liquid chromatographs, mass spectrometers, and liquid scintillation counting devices. One or more conventional fume hoods may be present when hazardous chemical procedures that do not require a sterile air environment are carried out in connection with the biological work.

12.1.4 Exclusions

Biosafety laboratories are not usually designed for work with greater than trace amounts of radioisotopes, carcinogens, mutagens, teratogens, or highly toxic systemic poisons. Restrictions on quantities of flammable solvents follow very closely the safe handling practices applicable to analytical chemistry and similar laboratories. There is rarely, if ever, a need for high-voltage and high-current electrical services, or for radiofrequency generators and other electrical operations with a high potential for fire, explosion, or electrocution. Other high-energy-releasing sources, such as lasers, X-rays, and gamma energy emitters that are lethal to biological agents, are unlikely to be present in the biosafety laboratory.

12.2 LABORATORY LAYOUT

12.2.1 Introduction

A biosafety laboratory may be a single room amid other laboratories of divergent uses or it may be a suite of many rooms interconnected through pressurized central corridors with building entrances and essential common service facilities, for example, animal quarters, supply rooms, and washing and sterilizing services. The minimum dimensions of a biosafety laboratory containing one Class II biosafety cabinet with auxiliary equipment and suitable for a single worker, is 7.5×10 ft. It should include, in addition to a 4- or 6-ft-wide cabinet, washing facilities, autoclave, and a space for donning and discarding protective garments at entry. Cloakroom and lockers should be provided for personal articles not required in the laboratory. A typical layout of a biosafety laboratory is shown in Figure 12-1. The recommendations contained in Sections 1.2 and 2.2 apply to biosafety laboratories and should be followed except as supplemented or modified in the following sections.

12.2.2 Floors and Walls

Biosafety laboratories require impervious surfaces and structural joints that are vermin-proof and easily cleaned and decontaminated. Walls and floors should be monolithic and made of washable and chemically resistant plastic, baked enamel, epoxy, or polyester coatings. The monolithic floor covering should be carried up the wall base with a smooth cove joint.

12.2.3 Access Restrictions

Access to biosafety level 3 laboratories should be limited by providing self-closing and self-locking secure doors all around, plus a double-door change room and shower facility at the entry of the laboratory. All self-

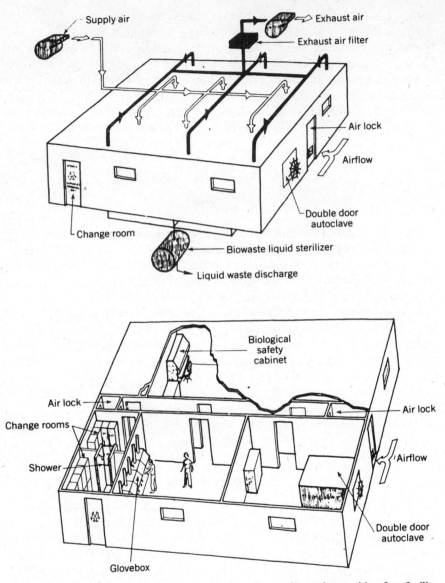

FIGURE 12-1. Secondary barriers in a representative total-containment biosafety facility.

CDC – SAFETY OFFICE   **FIGURE 12-2.** Biohazard sign.

closing and self-locking doors must present no barrier to egress in the event of an accident inside the laboratory. The biosafety laboratory should be maintained under negative pressure relative to corridors, offices, and so on (in all cases with nonrecirculated air), as an aid in preventing release of biological agents to areas outside the laboratory. All handling of hazardous biological agents should take place in biological safety cabinets.

Because access to biosafety level 3 laboratories is severely restricted, it is helpful to install windows consistent with the fire rating of the wall. Ease of surveillance of laboratory operations promotes safety. Systems for voice communication between laboratory personnel and others outside the laboratory are recommended.

12.2.4 Handwashing Facilities

Each laboratory room should contain one or more foot- or elbow-operated handwashing sinks.

12.2.5 Decontamination

Each laboratory or laboratory suite must contain an autoclave for decontaminating wastes.

12.3 HEATING, VENTILATING, AND AIR CONDITIONING

12.3.1 Introduction

Biosafety laboratories containing several biological safety cabinets that must be vented to the roof often require complex HVAC facilities and sophisticated controls to maintain adequate air supplies and the correct pressure relationship among adjacent spaces.

The HVAC comfort requirements for a biosafety laboratory with regard to temperature, humidity, and minimum air exchange rates are those outlined in Sections 1.3 and 2.3.

12.3.2 Ventilation

Class II, Type A biosafety cabinets are designed for exclusive use with biological hazards and may discharge all the exhaust air to the laboratory after filtration through a HEPA filter. Class II, Type B, biosafety cabinets are designed for use when treating hazardous biological agents with small quantities of radioactive tracers and carcinogenic chemicals. They are designed to discharge all exhaust air to the atmosphere through a dedicated exhaust system that terminates above the roof of the laboratory building. Therefore, the number and type of all fume hoods and Class B biosafety cabinets must be identified early in the design stage so that adequate supply air can be provided to satisfy exhaust requirements and maintain predetermined pressure gradients.

Practice varies with regard to shutting down airflow in biosafety cabinets. If they are shut down during non-working hours, the HVAC system must be carefully modulated to reduce supply and exhaust airflows while maintaining design room air pressure gradients. This requires sensitive pressure sensors and rapid-response feedback mechanisms, as well as visible gages for monitoring correct functioning.

12.3.3 Filtration

For some classes of agents, all exhaust air from biological safety cabinets, even when discharged directly to the atmosphere, must be filtered through HEPA filters for environmental protection. When using trace quantities of especially toxic volatile chemicals in conjunction with biological experiments, it may be prudent to add an efficient adsorber (generally activated carbon) as a second effluent air cleaning stage.

12.3.4 Controls

Pressure control can be maintained within a biosafety laboratory by providing a constant ratio of supply to return and exhaust air with the aid of differential pressure controllers, modulating dampers, fan inlet vanes, or a combination of all of these. Airflow variations should be minimized as an aid in providing control over room pressure. Continuous operation of hoods, cabinets, and local exhaust facilities is an important aid in maintaining reliable pressure control at all times. Air locks to adjacent areas should be provided. The specifications for air locks should be the same as those required by the General Services Administration in Federal Standard 209b (GSA, 1979).

12.3.5 Alarms

Alarms should be provided for the HVAC systems in a biosafety level 3 and level 4 laboratory. An additional alarm should be provided to notify personnel when HEPA filters are becoming dust loaded to a critical point. The information on air cleaning system monitoring instruments and alarms contained in Section 8.3.2.2 and 8.3.2.4 for the clean room laboratory also apply to biosafety laboratories.

12.4 LOSS PREVENTION, INDUSTRIAL HYGIENE, AND PERSONAL SAFETY

All provisions in Sections 1.4 and 2.4 apply to biosafety laboratories and should be implemented.

12.5 SPECIAL REQUIREMENTS

12.5.1 Warning Signs

Proper identification of hazardous biological agents is necessary to restrict traffic into hazardous areas and to alert all who enter the area to take precautionary measures. A standardized, easily recognized sign is customarily used for this purpose. It is displayed at each entry to the restricted area at a place where it can be seen easily and is displayed *only* for the purpose of signifying the presence of actual or potential biological hazards. Because entry control is very important for biosafety laboratories, an internationally recognized biohazard warning sign, colored magenta, has been adopted. It is shown as Figure 12-2.

12.5.2 *Personal Protective Equipment*

Personal protective equipment, such as laboratory coats, gloves, and disposable masks, should be issued to and worn by all who enter a biosafety level 3 laboratory. In some cases, shoe covers, head covers, and coveralls will be also required. Selection of adequate personal protective equipment is the responsibility of the laboratory director, but provision for storage of clean garments and safe disposal of those that become soiled must be delineated at the laboratory design stage.

12.5.3 *Decontamination*

It is frequently necessary to decontaminate biological safety cabinets with gaseous formaldehyde and sometimes necessary to decontaminate entire laboratories by this method. To decontaminate an entire laboratory with gaseous formaldehyde it is necessary to isolate the space in a gas-tight condition. Although temporary plastic curtains and caulking compounds may have to be resorted to in all cases to achieve ultimate leak tightness, forethought in the design of biosafety laboratories can simplify the job enormously with an equivalent saving in expense and down time. An example of constructions to be avoided in biosafety laboratories is the conventional drop ceiling with interconnections to other laboratories and offices in the space above the laid-in ceiling panels.

Formaldehyde is a suspected human carcinogen but no satisfactory substitute for decontaminating cabinets and laboratory rooms is known. Scrupulous attention to correct safety practices and the faithful use of personal protective equipment are essential when decontaminating with formaldehyde.

13

Clinical Laboratory

13.1 DESCRIPTION

13.1.1 Introduction

A hospital laboratory provides all of the routine clinical testing required for patient care. In-patient and out-patient specimens are collected, tests are conducted, residual specimens and completed test materials of a chemical and biological nature are disposed of in a safe manner, and reports provided.

13.1.2 Work Activities

Clinical laboratory activities include common procedures associated with hematology, bacteriology, and pharmacology. They involve mixing, blending, centrifuging, heating, cooling, distilling, evaporating, diluting, plating-out pathogens, examining specimens under the microscope, making radiochemical measurements, plus many similar operations. The use of automated, electronically controlled instruments to perform routine tests in busy hospital laboratories has become prevalent, thereby reducing to a minimum the need for handling chemicals. Clinical microbiological techniques have not advanced in a similar way and they remain largely manual operations.

13.1.3 Materials and Equipment Used

The amount and variety of materials and equipment that will be present in a clinical laboratory depend upon the type of facility it serves, that is, whether it is a large general teaching hospital, a satellite medical center, a clinic, or a doctor's office. Typical equipment includes microscopes, hotplates, mixers, autoclaves, balances, centrifuges, and such special instruments as blood cell counters, atomic absorption spectrometers, gas and liquid chromatographs, and mechanized, automatic specimen-analyzing devices of a chemical nature. Because many manual operations are still performed, handling of presumptive infectious specimens and examination of bacterial, viral, and fungal cultures derived from infectious specimens should be conducted in biological safety cabinets. See Section 12.1.3 and Appendix Vf for information on biological safety cabinets.

13.1.4 Exclusions

Hospital clinical laboratories are not designed for handling large quantities of hazardous materials or for conducting dangerous operations; in fact, neither should occur in a medical care facility. The procedures used in hospital laboratories are largely standard and routine. Clinical laboratories seldom engage in research for its own sake or conduct unusual test procedures, although some types of medical research utilize routine clinical test results as essential elements of data. There are no unusual access restrictions for a clinical laboratory.

13.2 LABORATORY LAYOUT

13.2.1 Introduction

The layout of a clinical laboratory will be determined by its size and the nature and number of clinical tests that will be performed. It usually resembles a combination analytical chemical and biological laboratory. Good practice standards for laboratory layouts, as outlined in Sections 1.2 and 2.2, should be followed. Special requirements may be imposed by The Joint Committee on the Accreditation of Hospitals (JCAH, 1987).

13.3 HEATING, VENTILATING, AND AIR CONDITIONING

The HVAC system for a hospital clinical laboratory will be required to maintain reasonable temperature control to ensure correct operation of the various electronic and other testing devices that will normally be present. The recommendations contained in Sections 1.3 and 2.3 are generally applicable to hospital clinical laboratories and should be considered for implementation with the following additions and comments.

13.3.1 Chemical Fume Hood

Chemical fume hoods are seldom needed in clinical laboratories, but when they are used, they should conform to the recommendations contained in Section 2.3.4.3. and Appendix V. Perchloric acid hoods are required in this type of laboratory only when perchloric acid digestions of samples are conducted routinely. When perchloric acid hoods are needed, the recommendations contained in Section 2.3.4.3.3 and Appendix Vc should be followed.

13.3.2 Local Exhaust Air Facilities

When there are discrete systems or processes in the hospital clinical laboratory that emit dangerous or obnoxious fumes (e.g., some hematology procedures) or large amounts of heat (e.g., autoclaves), it is advantageous to reduce the total ventilation air requirements for the laboratory by providing local exhaust air facilities for each device rather than to depend upon the general ventilation system for this purpose. Local exhaust facilities may consist of a canopy hood directly over the process or equipment, an engineered slot-type capture hood especially designed and built for the application, or a simple flexible exhaust hose to handle a small emission source such as an atomic absorption instrument flame. See Appendix Vg for more information.

13.3.3 Biological Safety

Biological safety cabinets are needed in many hospital clinical laboratories to handle specimens from infectious patients. Biosafety cabinets should be ventilated in accordance with requirements contained in Section 12.1.3. Most Class II biological safety cabinets used in hospital clinical laboratories will be Type A models that permit recirculation of air inside the laboratory, rather than requiring a direct exhaust connection to the roof. If volatile chemotherapy drugs will be associated with clinical specimens, it is advisable to use a Type B cabinet that is exhausted to the outdoors.

13.3.4 System Components

Hospital clinical laboratories are generally served by a central building HVAC system. Accepted practices of ASHRAE (ASHRAE, 1982, 1983,

1984, 1985, 1986) and SMACNA (SMACNA, 1985) are satisfactory and no special treatment of system components is required.

13.3.4.1 Supply Air Systems

Hospital clinical laboratories are acquiring increasing amounts of large, bulky, automatic electronic analytical equipment and, as a consequence, distribution of supply air is becoming more difficult, not only to maintain a uniform temperature in the room, but to prevent drafts and unsatisfactory air distribution patterns. For best results, supply air should be provided by a ducted distribution system located above the ceiling that discharges to the laboratory through multiple outlets designed to avoid drafts.

13.3.5 Temperature Control

A uniform temperature ($\pm 1.5°F$) is needed for reliable operation of some analytical devices. Heating and air conditioning control systems must be provided. When close humidity control is required, although it is seldom necessary, simultaneous operation of heating and cooling systems may be needed. Many kinds of HVAC systems, including local cooling units and central systems, have been used for hospital clinical laboratories with success. An important consideration is that all elements of temperature control systems should be interlocked so that a uniform temperature can be maintained throughout the year. Many times it is advantageous to install a separate air cooling and heating system for a hospital clinical laboratory rather than connect it to the building system because the laboratories have HVAC requirements that are different from those of the rest of the hospital building. When not separated, a large hospital central system will have to be operated at an inappropriate, and hence an uneconomical, level to maintain desired conditions for a small laboratory suite.

13.4 LOSS PREVENTION, INDUSTRIAL HYGIENE, AND PERSONAL SAFETY

13.4.1 Introduction

The recommendations contained in Sections 1.4 and 2.4 are generally applicable to clinical laboratories and should be considered for implementation with the following additions and comments.

13.4.2 Egress

Because clinical laboratory activities range from simple to very complex and diverse, segregation of activities should be considered in large facilities. Otherwise, maintenance of adequate exit routes and routes to fire extinguishers, deluge showers, and other emergency equipment may become difficult.

Chemical storage facilities should be carefully located with exit access, fire fighting, spill control, and nonrelated laboratory operation exposures in mind. Storage within the laboratory should be provided only for small amounts of chemicals, and this restriction should be maintained by careful laboratory management.

13.4.3 Fire Suppression

When placing hand-portable fire extinguishers in the laboratory, the requirements of Section 1.4.4.2.2 should be tempered with the fact that in a hospital facility it may be more critical to get to a fire extinguisher than to be able to make a rapid exit. For this reason, 2A-40 BC dry chemical extinguishers, with their increased extinguishing capacity over CO_2, are considered the most appropriate for use in this type of laboratory. They should be placed in the back of the laboratory as well as at each exit door to enable personnel remote from the exit to get to a unit quickly and safely.

13.4.4 Codes

NFPA 56C (NFPA 56C, 1985), which covers laboratories in health-related institutions, should be consulted for additional regulatory requirements.

14

Teaching Laboratory

14.1 DESCRIPTION

14.1.1 Introduction

A teaching laboratory should be planned, designed, and constructed to provide a safe working and learning area for groups of students. In high schools, the number of students may not exceed 30, whereas in most undergraduate college and university laboratories, the number is sometimes larger. Graduate-level laboratory instruction is normally conducted in research laboratories, which are covered elsewhere in this book. Teaching laboratories should be designed to demonstrate and encourage safe practices and operations because a disregard or ignorance of safety at the student stage will be carried over into the professional work that follows schooling. For example, laboratories designed for physics that involve electrical apparatus capable of providing serious electrical shock hazards or an ignition source potential should not be combined with chemical laboratories that use liquid chemicals or flammable liquids and gases.

14.1.2 Work Activities

Tasks performed in teaching laboratories will fall into two laboratory types: wet laboratories and dry laboratories. Wet laboratories employ bench experiments that use liquid, solid, and gaseous chemicals, heating devices, and open flames, and that discharge both gaseous and liquid effluents.

Dry laboratories use few liquid chemicals. They are characteristic of traditional physics, engineering, and biology laboratories. Experimentation involves the use of electrical components, light generators and optical instruments, mechanical devices, and microscopes, with some use of city gas and water.

14.1.3 Equipment and Materials Used

The materials and equipment found in teaching laboratories are determined by the subjects that are taught. A general chemistry teaching laboratory, for example, will tend to resemble a general chemistry research or analytical chemistry laboratory with respect to equipment and layout, although a teaching laboratory will have its own unique features.

14.1.4 Exclusions

Teaching laboratories, as defined here, are not intended for use for specialized research activities or for conducting hydraulics, civil engineering, materials testing, mechanical engineering, or electronics work. These laboratory types have unique requirements and more closely resemble pilot plants.

14.2 LABORATORY LAYOUT

14.2.1 Introduction

Teaching laboratories, wet or dry, require a maximum number of work stations in a minimum area. In spite of the pressure to maximize use of all available space, benches should be so located that easy, multidirectional movement and egress are maintained. Ease of movement is needed to facilitate getting to and from supply points or rooms. In addition, instructors must be able to move about freely, to see all areas, and to provide quick response to emergency situations. Peninsula arrangements do not permit such movement but wall benches and island benches do. Island-type benches for teaching laboratories are recommended.

The minimum distances discussed in Section 2.2.2.2 should be maintained between benches and between benches and walls. Experience shows that 32 ft^2 per student is an absolute minimum for a teaching laboratory and this minimum should only be considered when other aspects of the design are such as to allow ideal placement of fume hoods, adequate

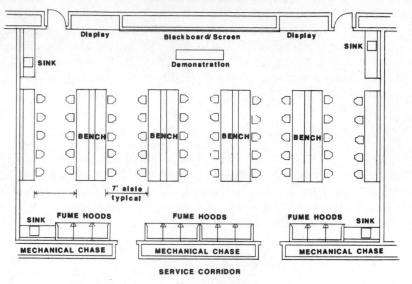

FIGURE 14-1. Teaching laboratory: Sample layout.

circulation when the room is fully occupied, and rapid and easy egress in case of emergency. By the time floor space per student reaches 70 ft^2 there is adequate room for design flexibility.

Distance between benches when students must work back to back must not be less than 6 ft. Otherwise, safe circulation is not possible for students and teachers who might be carrying chemicals, equipment, or other materials. Figure 14-1 shows a typical arrangement for a teaching laboratory.

With the increased use of microchemistry analytical techniques careful consideration should be given to the individual student work stations. For example, work stations may be designed for the student to be seated rather than standing because greater manual precision is needed.

14.2.2 Egress

There should be a minimum of two exits from each teaching laboratory, with each exit opening into a separate fire-safe egress. When teaching laboratories are large, additional exits may be required to be certain that a travel distance of 50 ft to an exit is never exceeded. All exit doors should swing in the direction of exit travel.

All of the remaining egress recommendations in Chapters 1 and 2 should be followed.

14.3 HEATING, VENTILATING, AND AIR CONDITIONING

The HVAC recommendations contained in Chapters 1 and 2 are generally applicable to teaching laboratories and should be considered for implementation. Additional comments regarding HVAC facilities for teaching laboratories follow.

14.3.1 Chemical Fume Hoods and Other Local Exhaust Systems

Fume hoods should be so located that they are remote from the main entrance and exit, and do not face exit routes or block them in the event of a fire, explosion, or a violent reaction within the hood. Hoods should be located near the back or outer walls of the laboratory, near the least-traveled pedestrian routes, but within easy access to the students.

At least one local exhaust facility is desirable at each bench in a wet laboratory to provide a convenient location where effluents from small fuming, smoking, or noxious experiments can be removed safely. We recommend one individual local exhaust facility per student in organic chemistry teaching laboratories but only one per four work sites for general chemistry teaching laboratories. Bench exhaust facilities are not usually needed for physics and similar teaching laboratories, but at least one chemical fume hood should be provided in every laboratory for solvent dispensing and dry activities that should not be conducted on an open bench. Downdraft exhaust ventilation should be considered where microanalytical techniques are used. Examples of local exhaust ventilation are presented in Appendix Vg.

14.4 LOSS PREVENTION, INDUSTRIAL HYGIENE, AND PERSONAL SAFETY

The loss prevention, industrial hygiene, and personal safety recommendations contained in Chapters 1 and 2 are applicable to teaching laboratories and should be considered for implementation in addition to the recommendations that follow.

14.4.1 Emergency Showers

Emergency showers for large teaching laboratories should be placed inside the laboratory proper and so located that no more than 25 ft of travel distance is required from any point. Showers should not be placed in front of chemical or flammable liquid storage cabinets or shelves, or directly in front of fume hoods. Additional requirements for emergency showers are given in Appendix VII.

14.4.2 Emergency Eyewash

For laboratories using chemicals and containing four or more multistudent benches, an eyewash facility should be located at each bench. This can be a hand-held type. For laboratories with fewer than four benches, there should be at least one eyewash facility per laboratory and it should be so located that no more than 3–4 s is required to reach it from the most remote work station. In addition to the hand-held type eyewash mentioned above, at least one tempered full-face eyewash facility should be located in each teaching laboratory. Additional information on eye wash facilities is given in Appendix VIII and in Section 2.4.1.5.

14.4.3 Chemical Storage and Handling

No highly reactive or flammable chemicals should be stored in a teaching laboratory. An adjacent chemical storage room or a specially constructed protected area should be provided for this purpose. In teaching laboratories, provisions should be made to shelve or otherwise hold only the amount of chemicals necessary for a day's experiments. Construction details of safe chemical shelving and storage cabinets are covered in Section 2.4.

14.4.4 Hazardous Chemical Disposal

Central points for the collection and temporary storage of chemical waste should be provided in each laboratory. They should be remote from students at their work sites and not located in egress routes. Locations near fume hoods are recommended. Large waste storage areas are unnecessary because wastes should be removed at least daily.

14.4.5 Fire Extinguishers

Provisions should be made for locating fire extinguishers within each teaching laboratory. With island benches, one fire extinguisher should be located at each bench. The type of extinguisher is dependent on the use of the laboratory. A clean agent such a CO_2 is appropriate for chemical operations. Size 4A-40 BC or larger ABC-type dry chemical units should be located in the hall to be used as a backup.

14.5 SPECIAL REQUIREMENTS

14.5.1 Preparation Room

There should be a room associated with each teaching laboratory, or group of teaching laboratories, that can be used for the preparation of experimental equipment and materials. If the teaching laboratories do not involve the use of chemicals or hazardous substances, no special facilities are needed. When chemicals are used, the following considerations are important.

Chemicals stored in the preparation room (when not stored in the areas referred to in Section 14.4.3) will be in the nature of bulk chemical storage. Approved storage cabinets should be provided (see Sections 1.4.7 and 2.4.6) in adequate numbers to handle all flammable liquids. Provisions should be made to store all chemicals according to safe compatibility characteristics. The preparation room should have good general ventilation to dilute released materials below their hazard level, that is, explosivity in the case of flammables and toxicity in the case of toxic materials.

The preparation room should have a well planned fire-suppression capability that includes fixed automatic fire-suppression facilities and hand-portable fire extinguishers of the ABC dry chemical type with ratings of 4A-40 BC or better. Hand-portable extinguishers should be located strategically to assure that a fire can be attacked quickly and kept from threatening the laboratory itself.

An arrangement such as a pass-through or a Dutch door should be used to eliminate the need for students to enter the preparation room.

15

Gross Anatomy Laboratory

15.1 DESCRIPTION

15.1.1 Introduction

The gross anatomy laboratory is a group of spaces for the preparation, storage, and dissection of human cadavers, animal bodies, or portions thereof, for the purpose of teaching gross anatomy or for research. The spaces described in this section include the receiving area, morgue, cold storage room, and dissection room. Emphasis will be given to the gross anatomy teaching laboratory because of its special requirements beyond those of anatomical research.

15.1.2 Work Activities

15.1.2.1 Morgue

Medical research and teaching institutions gain anatomical gifts from persons who arrange before death for the donation of their bodies after death. The deceased are brought to the morgue receiving dock by morticians or others. The donated bodies are prepared in a manner to preserve all tissues, so that the cadaver can remain unrefrigerated for long periods without harmful deterioration. The cadavers are pickled, then placed either in cold boxes for short-term storage, or in freezers for long-term storage. After dissection or use, all of the remains are returned to the morgue to be transported for burial or cremation.

15.1.2.2 Dissection Laboratory

Gross anatomy is generally taught by lectures in conjunction with laboratory sessions. In the laboratory, students inspect and identify all tissues revealed by dissection. Dissections are done by students, but may be done by the instructors. The activities include cutting, sawing, manipulation with fine instruments, and cleaning up. During laboratory sessions, instructors may give demonstrations with models and visual aids, with or without previously dissected specimens called prosections.

A single cadaver may be dissected by a group of 4 to 6 students for the entire course, which may be two weeks to a semester. Students refer to their anatomy textbooks, adjacent to the dissection table, as they manipulate and view the cadaver. Both visual and tactile information is gathered. Students must be able to inspect the cadaver from all angles and at very close range, 6 in. to 1 ft away. At many schools, students are required to clean up the dissection rooms.

15.1.2.3 Research Laboratory

The activities in research laboratories that use gross anatomical specimens may include cutting, mixing, mechanical or biochemical procedures, tissue culture, microscopy, and tissue sample preparation and staining.

15.1.3 Equipment and Materials Used

15.1.3.1 Morgue

The morgue contains mortuary equipment for the removal of blood, body cleansing, and perfusion of the bodies with preservatives. Chemicals used include glycerin, glycol, formalin (formaldehyde), phenol (carbolic acid), soap, and water. Organic pathologic materials may be present. Personnel must be trained to recognize and safely handle these specimens. Saws and other instruments are used to collect body parts for separate storage in jars or for prosections. There are large scales and materials-handling equipment, which may include carts, gurneys, and forklift trucks.

All surfaces within the morgue should be impermeable, nonporous, and washable. Formaldehyde vapor permeates all paper, cloth, plastic, and wood materials stored in the morgue. Permeable supplies should be stored in a separate room. Bulk chemicals should also be stored separately. Floors should be covered with a seamless, smooth material, with an integral cove that extends 8 in. up each wall. Flooring that becomes slippery when wet should not be used. Flooring that may crack under the loading of a forklift truck should be avoided. All drains in a morgue should have grease traps, so that organic matter can be cleaned out of the waste water.

Walls should be waterproof, seamless, and smooth. Glazed tile or epoxy paint on block or water-resistant sheetrock are recommended.

Plaster or concrete ceilings can be painted with epoxy. Suspended ceilings may be lay-in types of metal pans, preferably nonperforated. Light fixtures, wall outlets, and switches should be vaporproof because of the moist environment brought about by frequent cleaning procedures.

15.1.3.2 Dissection Laboratory

The materials present in the gross anatomy dissection laboratory may include preserved bodies, unpreserved anatomical specimens, organic waste matter, and chemicals. Chemicals include glycerin, glycol, formalin (formaldehyde), phenol (carbolic acid), soap, and water. Equipment, in addition to dissection instruments, may include saws, drills, and cleaning tools. There will be dissection tables which may be mobile or fixed. Teaching aids may include models, skeletons, charts, and audiovisual equipment.

Formaldehyde, recognizable at concentrations as low as one part per million in air, will be ever-present. Therefore, all surfaces should be impermeable, nonporous, and smooth, as in the morgue. Even more than in the morgue, surfaces here should be easily cleanable, since students may have to do the work. The recommended material for all furnishings in a dissection laboratory is stainless steel. All metal joints should be welded and coved for thorough cleaning of tables, sinks, countertops, and storage cabinets. Wood and plastic materials should be avoided for construction or in furnishings.

For ceilings in gross anatomy dissection laboratories, acoustical qualities may be more important than impermeability. The noise level during classroom hours could make a dissection laboratory with all hard surfaces unbearable.

All floor and sink drains should have grease traps for capture of organic matter, although cleaning of these traps may cause a service personnel problem. A means of mechanizing this chore should receive serious consideration in the design planning phase.

15.1.3.3 Research Laboratory

The materials present in a research laboratory that uses gross anatomical specimens are those listed above for the dissection laboratory. In addition, all the chemicals, equipment, and instruments associated with a general research laboratory may be present.

15.1.4 Exclusions

15.1.4.1 Morgue

An educational or research institution morgue should not be used for clinical autopsies or commercial use. The morgue should be restricted from public access, including donors' family members and students. Authorized, trained technicians and faculty are charged with operating the morgue to maintain the privacy and dignity of those who make the donations.

15.1.4.2 Dissection Laboratory

The dissection laboratory is not an operating room or laboratory for experimentation with live animals. The dissection laboratory's electrical and ventilating systems are not designed for use of anesthetics.

15.2 LABORATORY LAYOUT

15.2.1 Introduction

The public may regard the activities pursued in morgues and dissection laboratories with alarm, disgust, or morbid curiosity. To secure these areas from public access and view, the location of the morgue and dissection laboratories should be carefully considered so as to provide for the following:

Access to a loading dock and receiving room.
Private passageway between receiving room and the morgue.
Private passageway between morgue and dissection laboratories.
Windows not overlooked by other buildings.
Entries to morgue and dissection laboratories not on a main corridor.

There should be a covered receiving dock and enclosed room that contains a mortuary-type refrigerator or cold box for the deposit and temporary storage of bodies brought by a mortician or others when the morgue is unattended. The cold storage facility should be large enough for donations made over long holiday weekends. Access from this room to the morgue should be restricted. Refrigerators or cold boxes that can be opened from either side permit discrete transfer of donations. The receiving room should be adjacent to a morgue on the same level or connected by a key-restricted elevator to a morgue on a different level. Similar private circulation is desirable between the morgue and the dissection rooms to reduce the chance of accidental contact with the public.

15.2.2 Individual Room Arrangements

15.2.2.1 Morgue

The layout of a morgue for a research or medical education institution depends on the projected quantity of anatomical gifts, number of staff, and whether there is an anatomical collection or museum. The morgue consists of a preparation room, materials and chemical storerooms, pickling rooms, cold rooms or mortuary refrigerators, sample preparation rooms, and locker rooms. There should be space in the preparation room for as many gurneys as needed to transport all the donations held in the receiving area. Adequate space is needed to make it easy to maneuver the heavily loaded carts from one procedure area to the next. Bodies are moved temporarily to racks in room-temperature storage areas immediately adjacent to the preparation area. The alignment of racks and aisles should be arranged so that the staff can safely move the cadavers into cold rooms or walk-in freezers adjacent to the morgue or dissection laboratories. In the cold rooms, cadaver storage can be horizontal or vertical. Every area in the morgue should be arranged to facilitate transport.

The support areas for materials and chemical storage, sample preparation, offices, and locker and shower facilities should be immediately adjacent to the preparation room. These functions are kept together so that the formaldehyde odor can be contained.

15.2.2.2 Dissection Laboratory

A gross anatomy teaching laboratory consists of locker/shower rooms for men and women, dissection rooms, and storerooms. These functions should be kept together for security and odor containment. Seminar rooms may also be adjacent to the laboratory suite. Circulation need not to be in a one-way loop, as in a surgical area. Dissection tables are 2 ft 11 in. high, 6 ft 6 in. long, and 2 ft 6 in. wide. They are usually arranged with minimum clearances of 50 to 54 in. head to toe and 40 to 50 in. side to side to accomodate an average of four students per cadaver plus their books and instruments. More area per table is required if students in wheelchairs are attending. There should be good light and ventilation at all tables. Large scrub sinks should be provided for handwashing and general cleanup. Countertops and display cases can be distributed along the perimeter of the room for additional teaching materials. Skeletons require hangers or racks, but all should be stored in locked cabinets or storerooms. Blackboards and projection screens should be positioned so that all students can see them.

15.2.2.3 Research Laboratory

The layout recommendations contained in Chapters 1 and 2 are generally applicable to research laboratories that use gross anatomical specimens and they should be followed closely.

15.2.3 Egress

The egress recommendations contained in Chapters 1 and 2 are generally applicable to gross anatomy laboratories and should be followed closely, in addition to the following recommendation.

15.2.3.1 Morgue

There should be two exits from the morgue preparation area. Exits from cold boxes and walk-in freezers should be designed with the egress recommendations contained in Chapters 1 and 2 firmly in mind. All doors should be operable from either side.

15.3 HEATING, VENTILATING, AND AIR CONDITIONING

15.3.1 Introduction

HVAC systems for anatomy laboratories are critical because they must provide a safe environment for those who must work in a laboratory with potentially high concentrations of formaldehyde and other toxic chemicals. The HVAC recommendations contained in Chapters 1 and 2 are generally applicable to gross anatomy laboratories and should be followed closely. The basic intent is to maintain formaldehyde, phenol, and other chemical concentrations below objectionable levels. Odor control in the laboratory and surrounding area is also critical. Additional recommendations follow.

An outside air change rate of 15 to 20 per hour is recommended for laboratory and morgue. It is best to introduce the supply air high in the room and to exhaust it from a low room location to draw the contaminated air below the work area and well below the breathing zone. Specially equipped fixed autopsy tables with built-in downdraft exhaust could provide local capture of chemical fumes more effectively than general room exhaust points located near the floor.

Recirculation of air is not recommended. Certain chemicals, such as potassium permanganate, that have good absorption/adsorption proper-

ties for formaldehyde have been used in some cases to remove formaldehyde vapors from recirculated air. For such systems to be successful,
large recirculation rates must be maintained and careful monitoring of the
activity of the air cleaning chemicals must be conducted on a continuing
basis.

Local exhaust or capture hoods may not be feasible for gross anatomy
teaching laboratories because students tend to work extremely close to
the parts being dissected. Therefore, low room air exhaust grilles located
on the perimeter walls are likely to be the most effective solution to
embalming fluid vapor exposure. Nevertheless, local exhaust opportunities should be studied for each function to determine the feasibility of this
method of vapor and odor control. Use of stainless steel ventilation ducts
should be considered because of their corrosion resistance. The room
should be at a negative pressure with respect to all public areas. No
special treatment of exhaust air is needed provided the room exhaust air
can be discharged well above all surrounding buildings and terrain features.

15.3.2 Individual Room Requirements

15.3.2.1 Morgue

Cadavers are usually prepared before they are received in the morgue or
storage area. However, some are prepared in the morgue. The embalming
process consumes substantial amounts of formaldehyde and exposure
levels are necessarily higher when embalming occurs in the morgue.

Automatic feed systems into closed vented vessels should be used to
prepare large volumes of embalming fluid. Small quantities can be prepared in a laboratory fume hood or ventilated enclosure.

15.3.2.2 Dissection Laboratory

In gross anatomy teaching laboratories, students will be dealing with
cadavers already prepared and preserve in embalming fluid, an aqueous
solution of formaldehyde, methanol, ethanol, phenol, and glycerine. The
extent of the chemical exposure depends upon the type of dissections
made and the organ being studied. For example, levels of exposure are
higher for dissections involving the opening of a body cavity. The greatest
exposure appears to be in the abdomen. Quantitative measurements sub-
stantiate this experience. The exposure levels sometimes do not decrease
with time because, as cavities dry out, students intermittently wet the
cadaver with additional embalming fluid.

15.4 LOSS PREVENTION, INDUSTRIAL HYGIENE, AND PERSONAL SAFETY

The loss prevention, industrial hygiene, and personal safety recommendations contained in Chapters 1 and 2 are generally applicable to gross anatomy laboratories and should be followed closely.

16

Pathology Laboratory

16.1 DESCRIPTION

16.1.1 Introduction

The pathology laboratory consists of a clinical or research facility plus support areas that include a clinical morgue, autopsy storage, organ storage, slide preparation and storage, and photography laboratory. The support areas are based within a hospital. The research laboratory may be at the same hospital or at another institution.

16.1.2 Work Activities

16.1.2.1 Clinical Morgue Autopsy Room

The performance of autopsies is the primary activity within a clinical morgue. Autopsy is legally required on persons who die of unknown, accidental, or suspicious causes. For the hospitalized individual, an autopsy is care for the patient that extends beyond death. It calls for cooperation between pathologist and clinician. An autopsy demonstrates the cause of death and condition of all organ systems by means of gross and microscopic inspection. In an autopsy, the pathologist collects samples of all organs to preserve for later investigation or research. The pathologist presents the results of the autopsy to the clinician for review, confirmation of diagnosis and treatment, and discussion.

Another activity conducted in clinical morgues is its use as an animal morgue associated with a clinic, research institution, or commercial animal-breeding facility.

16.1.2.2 Sample Preparation Room

Activities conducted in a sample preparation room include fixing whole organs and organ samples by immersion in sinks, tanks, or jars of formaldehyde and by freeze-drying. In addition, pickled and freeze-dried organ tissues are thinly sliced, fixed in chemical solutions, stained with a variety of dyes and chemicals, and then mounted on slides for microscopic examination.

16.1.2.3 Cold Storage Room

In a hospital morgue, bodies remain in cold storage for only a short time before they are autopsied or removed for funeral preparation by morticians. Family members may view the deceased before removal. The morgue may have a separate room adjacent to the cold room for this purpose.

16.1.2.4 Pathology Laboratory

Activities conducted in research or clinical pathology laboratories involve the use of diseased and damaged tissues plus contaminated or infectious materials of living organisms. Operations include tissue cutting, chemical mixing, mechanical manipulation of organs and tissues, biochemical procedures of widely varying nature, microscopy, microbiological culturing, sample preparation, and staining.

16.1.3 Materials and Equipment Used

16.1.3.1 Clinical Morgue Autopsy Room

Autopsy rooms contain autopsy tables, scales, instrument cabinets, tape recording equipment, tissue cleansing sinks, and handwashing sinks. An autoclave to sterilize contaminated or infected materials should be within or adjacent to the morgue. The autopsy room may also have an ultracold freezer for immediate preservation of tissues plus other refrigerators for storing donated organs. Chemicals used in an autopsy room include soap, water, alcohol, and decontamination agents.

All surfaces should be impermeable, nonporous, and washable so the morgue can be decontaminated. Flooring should be seamless and nonslippery, with an integral cove extending 8 in. up each wall. All drains should have strainers and grease traps so that organic matter can be cleaned out.

Walls should be waterproof, seamless, and smooth. Glazed tile or epoxy-painted concrete block or water-resistant sheetrock are recommended. All storage cabinets or cupboards should be stainless steel.

Plaster or concrete ceilings can be painted with epoxy. Suspended ceilings may be lay-in solid metal pan. Formaldehyde is not generally used in the autopsy room itself, so sound-attenuating material may be used above the metal pan ceiling. Due to the presence of moisture from frequent cleaning procedures, light fixtures, wall outlets, and switches should, preferrably, be vaporproof.

16.1.3.2 Sample Preparation Room

Equipment in a sample-preparation laboratory may include a cryostat, chemical fume hood, lyophilizer, microtome, and a large sink with a circulating pump for formaldehyde fixation. Contaminated materials and infectious organisms may be present. Tissues from patients with infectious diseases should be handled in biological safety cabinets specially designed for these agents (see Chapter 12 and Appendix Vf). All materials should be autoclaved before disposal to destroy harmful biological agents. Staff must be trained to handle infectious specimens safely.

Chemicals include formalin (formaldehyde), phenol (carbolic acid), soap, water, and various histological stains and fixatives.

All surfaces should be impermeable, nonporous, and washable, as in the autopsy room. Flooring should be seamless and nonslippery, with an integral cove extending 8 in. up each wall. All drains should have strainers and grease traps so that organic matter can be cleaned out.

Ceilings should not have sound attenuating materials because of the very high concentrations of formaldehyde that will be present.

16.1.3.3 Clinical Pathology Laboratory

A clinical pathology laboratory will contain contaminated or infectious materials as well as chemicals, equipment, and instruments associated with a general research laboratory.

16.1.4 Exclusions

16.1.4.1 Morgue

The clinical morgue should be restricted from public access and from hospital staff who do not work in the morgue or have related activities.

16.2 LABORATORY LAYOUT

16.2.1 Introduction

The clinical morgue and pathology laboratory are critical facilities in a hospital. Patients, visitors, and personnel should be reasonably shielded from these functions. As in the medical education morgue, the location of the clinical morgue is important. There should be a covered dock for receiving bodies for autopsy and for removing bodies for funeral preparation. This dock should not be shared with materials handling or trash removal activities. If the morgue is not adjacent to the dock, a short, private corridor or key-operated elevator should connect them directly without passage through other areas. For transport of patients who die in the hospital, personnel must use corridors used by the public and other staff. It is better for non-medical staff morale if the morgue is away from main service corridors so they are not exposed daily to corpses. The location of clinical pathology laboratories is not as critical. Tissue samples can be transported safely and discretely in containers. If highly infectious materials are frequently used, the laboratory should be adjacent to the source.

16.3 HEATING, VENTILATING, AND AIR CONDITIONING

16.3.1 Introduction

Environmental control in the pathology laboratory could be critical, depending upon the use of the facility. An important difference between a community hospital and a large teaching hospital is the much larger number of autopsies performed in the teaching hospital. During an autopsy, organs and tissue samples will be removed for study and examination. Unless they are immediately frozen at very low temperature, they will be preserved in formaldehyde solution. This usually results in very high concentrations of formaldehyde in the area. Patients who die with a serious communicable or infectious disease are usually autopsied. This presents a need to control an infection hazard and, in some hospitals, a separate area with strict access control and special facilities is designated for this service. HVAC systems must deal with each of these environmental control issues.

16.3.2 System Description

The information contained in Section 15.3.1 also applies to hospital pathology laboratories.

16.3.3 Local Exhaust Systems

Local exhaust ventilation hoods or capture hoods are strongly recommended for hospital autopsy rooms where large quantities of formaldehyde are used in open trays for transfer of tissues. Local capture hoods placed at work level are effective for controlling this formaldehyde exposure. Care must be taken in positioning such hoods to prevent obstruction of the procedure. A good location for them is near the sink area. Stainless steel ducts are recommended for corrosion control. Exhaust for autoclaves is strongly recommended. Canopy type hoods are commonly employed. They should be placed as close to the autoclave as possible.

16.3.4 Special Autopsy Requirements

For autopsy rooms dedicated to infectious cadavers, the use of stainless steel for ducts, air outlets, and all other equipment and facilities is desirable for its ease of cleaning. The room must be at a negative pressure with respect to its surroundings. No recirculation of air is permitted. It is strongly recommended that HEPA filters be placed in the exhaust air system. It is also necessary to provide facilities for paraformaldehyde decontamination of the entire room by adding shutoff dampers in the supply and exhaust ducts.

All of the precautions and safety regulations described for biosafety laboratories (Chapter 12) should be observed when handling an infectious cadaver.

16.3.5 Supply Air Systems

Supply air requirements will be similar to those for other laboratories. Air conditioning is highly desirable.

16.4 LOSS PREVENTION, INDUSTRIAL HYGIENE, AND PERSONAL SAFETY

The loss prevention, industrial hygiene, and personal safety recommendations contained in Chapters 1 and 2 are generally applicable to pathology laboratories and should be followed closely.

16.5 SPECIAL REQUIREMENTS

16.5.1 Waste Fluids

Disinfection of waste fluids from these facilities should be considered. Consult a safety engineer, industrial hygienist or infection control authority for assistance.

17

Team Research Laboratory

17.1 DESCRIPTION

17.1.1 Introduction

A team research laboratory usually occupies a single unpartitioned space larger than the average size of a two- to four-module laboratory. It is not defined by the specific activities conducted within. The term "team laboratory," as used here, is not interchangeable with the designation "laboratory unit." A laboratory unit is identified as an aggregation of laboratory-use spaces defined by the grade of fire separation between the individual spaces within the unit, as well as between the unit and abutting spaces. A team laboratory may also be a laboratory unit, but it need not be. The distinguishing feature of this laboratory type is the interdisciplinary nature of the activities conducted therein. The large size of a team laboratory often poses unique problems in layout, HVAC, and loss prevention.

17.1.2 Work Activities

Activities conducted in a team research laboratory include any or all those carried on in general chemistry, analytical chemistry, physics, clean room, controlled environment, radiation, and high-pressure laboratories.

17.1.3 Materials and Equipment Used

Equipment used in the team research laboratory may include any of the items listed as characteristic of the several laboratory types listed here in Part II.

17.1.4 Exclusions

Activities, materials, and equipment specifically associated with high-toxicity and biological hazard laboratories are not usually conducted in a team research laboratory.

17.2 LABORATORY LAYOUT

17.2.1 Introduction

The recommendations in Sections 1.2 and 2.2 apply generally to team research laboratories. The management of spills and other types of accidents involving hazardous materials is much more difficult in the large open area of a team laboratory, Figure 17-1, than in a room of limited size since there is no simple means of containing hazardous fumes or smoke. Because of this, there is increased risk of flame spread. An incident in one area of the laboratory may not be noticed in time for safeguards to be taken by those in other parts of the laboratory.

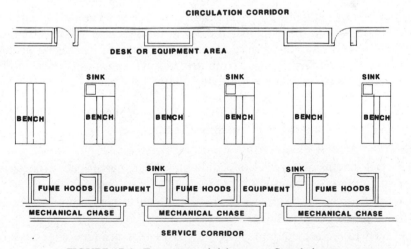

FIGURE 17-1. Team research laboratory: Sample layout.

17.2.1.1 Entire-Floor Laboratory

The team laboratory that occupies an entire floor of a building and is not divided by walls or egress corridors requires special planning. The space must be organized with a pattern of circulation that is easily perceived. Space allocation in the laboratory may vary from 60 to 200 ft^2 per person. It is important that the laboratory layout be sufficiently logical and simple that all personnel know where they are within the laboratory at all times and how to get out, even when vision is impaired by darkness, smoke, or chemical irritants. A rectangular grid is recommended for circulation because of its familiarity to all by sight or touch. With no visual clues, people can literally feel their way out. Whole groups can exit safely when the major aisles are wider than the standard aisles between benches and equipment, and this arrangement is recommended for all team research laboratories.

17.2.1.2 Utility Distribution

Utility outlets should be aligned with the circulation grid so that all arrangements of fixed and moveable benches and equipment will maintain no less than the minimum recommended aisle and egress passageway widths. Sinks should be evenly distributed so that all personnel are within 10 s travel time to an emergency eyewash station and for washing spilled materials off hands, and so forth.

17.2.1.3 Zoning

Because there may be a great variety of activities occurring simultaneously within a team research laboratory, well defined zones for particular activities should be established early in the planning phase to reduce conflicts. As in any laboratory, zones of activities of increasing hazard should be located farther away from primary egress aisles. Zones can be determined by observing the following characteristics:

Areas with fixed benches vs. open floor areas for moveable benches and large nonpermanent equipment setups.

Sterile processes vs. nonsterile processes.

Special exclusion areas for containment of hazardous processes, equipment, or materials.

A need for special exhaust air requirements to service processes or equipment that produce excessive heat or fumes.

A need to isolate equipment or processes that produce excessive or unpleasant noise.

17.2.2 Egress

The recommendations in Section 2.2.1 apply generally to team research laboratories. When laboratories are larger than four standard modules, the distance to exitways should not exceed 50 ft.

Egress markings should be in accordance with local and national standards. A special effort should be made to provide supplementary egress signs and visual guides for exit from large team research laboratories. This is particularly important when there are many exits. Signs should be visable from all areas and kept clear of pipes, ducts, light fixtures, and similar sight obstructions.

17.3 HEATING, VENTILATING, AND AIR CONDITIONING

The recommendations in Section 2.3 apply generally to team research laboratories.

17.3.1 Exhaust Air Systems

17.3.1.1 Fume Hoods

Large team research laboratories may require many chemical fume hoods. When they are grouped together, a lot of air must be exhausted from a limited area, causing undesirable drafts and turbulence in the airflow patterns. The use of auxiliary air chemical fume hoods, especially when they are placed in alcoves, reduces air current velocity and turbulence in the laboratory, and also reduces the total amount of conditioned building supply air that must be exhausted to the outside. See Appendix Vb for more information. The alcoves should be placed in high hazard zones away from the main egress aisles. In addition, the location of the chemical fume hoods must be carefully planned relative to the other local exhaust systems to avoid deflecting or reducing the effective capture velocities of these devices.

17.3.1.2 Local Exhaust Air Systems

Recommendations contained in Section 2.3 apply generally to team research laboratories.

17.3.1.3 General Room Air Exhaust Systems

Recommendations contained in Section 2.3.4.1 apply generally to team research laboratories. An even distribution of exhaust grilles is important to reduce the spread of odors and fumes throughout large team research laboratories.

17.3.2 Supply Air Systems

Recommendations contained in Section 2.3.3.1 apply generally to team research laboratories. An adequate volume and even distribution of supply air are especially important in large team research laboratories to provide for many exhaust air systems in a draft-free manner.

17.4 LOSS PREVENTION, INDUSTRIAL HYGIENE, AND PERSONAL SAFETY

The recommendations contained in Sections 1.4 and 2.4 apply generally to team research laboratories.

17.4.1 Emergency Showers

The recommendations contained in Section 2.4.1.4 apply generally to team research laboratories. The emergency showers in team research laboratories that cover an entire floor should be located within the laboratory rather than in the corridors to assure that there will be no more than 25 ft of travel from any point to a shower. The showers should be arranged so that splashing and water runoff will not endanger laboratory equipment, benches, or electrical panels. Appendix VII contains additional information on emergency showers.

17.4.2 Emergency Eyewash Facilities

The recommendations contained in Section 2.4.1.5 apply generally to team research laboratories. There should be one emergency eyewash station for every two modules, distributed evenly at sinks throughout the laboratory. In addition, there should be at least one eyewash facility of the full-face type with water tempered in accordance with recommendations in Appendix VIII.

17.4.3 Hand-Portable Fire Extinguishers

The recommendations contained in Section 1.4.4.2.2 apply generally to team research laboratories. In addition, large-capacity, multipurpose fire extinguishers may be needed within the laboratory. This should be determined with the aid of fire protection and safety specialists based on the anticipated uses of the laboratory.

18

Animal Research Laboratory

18.1 DESCRIPTION

18.1.1 Introduction

The design of large research facilities for housing and caring for major laboratory animal colonies, especially when provisions must be made for accommodating many different species in the same facility, is a complex task that should not be undertaken without the active advice and assistance of veterinarians and scientists experienced in animal research. The use of animals for research purposes and the facilities used for their care at universities and research institutions have come under intense criticism by a number of groups in recent years. As a result, new and more stringent regulations covering all aspects of animal care and animal research practices have proliferated, and it appears unlikely that this trend will be reversed in the foreseeable future. It is extremely important, therefore, that research facilities intended for animal housing and animal research be initially designed to meet the highest foreseeable standards. Anything less is likely to result in rapid obsolescence and to risk serious interference with future research programs.

Although the construction of large-animal housing and research facilities is beyond the scope of this manual, there is frequently a need for the provision of animal facilities of very modest size that are restricted to the housing of small numbers of only one or of a few similar species (e.g., only rodents). The animal laboratory described in this chapter is confined to this small-scale purpose and the information that is contained here should not be extrapolated to cover the design of larger-sized facilities or multispecies housing. The animal laboratory described here is suitable for the housing and care of the most-used small animals, principally rodents (mice, rats, guinea pigs, and rabbits). It is not designed for primates, dogs, fowl, or large animals such as sheep, donkeys, and pigs.

The minimum facilities required for a small unit research and teaching laboratory that utilizes small numbers of small animals, are the following:

1. New animal reception and quarantine area, usually a closed room isolated from the main animal quarters.
2. Animal holding rooms, where animals undergoing or awaiting experimentation are housed, fed, and cleaned.
3. Sanitation facilities, often designated as a "dirty area," to (a) wash and sterilize cages, water bottles, feed troughs, and so on, (b) collect and dispose of animal wastes and soiled bedding, and (c) dispose of dead animals. An in-house incinerator or steam sterilizer is needed when dealing with highly infectious human pathogens or potent carcinogens, but otherwise, commercial disposal facilities can be utilized.
4. Storage room with vermin-proof bins for animal feed, clean bedding, and other animal-care supplies.
5. Experimental treatment–surgery–autopsy room. When the size of the facilities permit, the autopsy room should be isolated from the treatment and study laboratory. Tissue preparation and pathology facilities can be provided external to the animal laboratory when space is limited.
6. Freezer for holding dead animals prior to disposal and excised tissues prior to examination.

In addition to purely humane considerations, good animal research cannot be conducted with sick, dirty, infested, and poorly tended animals.

Good animal laboratory design includes an efficient layout of facilities for ease of operations, plus selection of materials of construction that make it possible to maintain excellent sanitary conditions. Whatever else may have to be omitted from the animal research laboratory for reasons of inadequate space and insufficient budget, no compromise should be made in provisions for maintaining excellent sanitation throughout the facility. The important factors include the following:

1. Impervious monolithic floors, walls, and ceilings, constructed of materials and finishes resistant to cracking and damage from washing detergents and disinfectants, including chlorine-containing compounds.

2. Metal window frames that are resistant to damage from moisture and mildew and that can be permenantly closed to lock out pests.

3. Doors that are animal-proof when closed, to prevent entry of wild species (attracted by the easy availability of food) and loss of loose laboratory animals.

4. Careful sealing of all penetrations for utilities (electricity, water, drains, fire sprinklers, heating and ventilating ducts, etc.) to close off harboring places for vermin. It should be kept in mind that application of insecticides in active animal colonies is often prohibited because of the unknown influence these poisons may have on the outcome of experiments then underway. Therefore, it should be accepted from the start that the only useful vermin control program is strict prevention of infestation. This is one of the reasons why an isolated quarantine room is essential, to prevent the introduction into the main animal colony of infected and infested animals.

5. Some experts in pest control recommend applications of boric acid in wall spaces during construction or renovation in order to reduce cockroach problems.

Each animal species has an ideal air temperature, air humidity, and air movement rate that promotes health and longevity. Drastic average deviations and random substantial variations from these ideal conditions will reflect adversely on the longevity of the animals. The seriousness of uncontrollable environmental conditions within the animal laboratory will vary with the normal animal holding period. For acute toxicity experiments, as for determining LD_{50} data, longevity greater than a week is seldom important. However, for low-level chronic toxicity exposure experiments, extreme longevity of the experimental animals is critical to success, and every possible effort must be expended to promote the health and safety of the animal colony for the duration of long experimental periods.

Optimal spacing and HVAC conditions have been published for a number of commonly used animal species by the U.S. National Institutes of Health (NIH, 1985), as well as by counterpart agencies in other countries. Generally, for best results, it is necessary to maintain the interior laboratory climate with as little variation as possible; therefore, a windowless area is highly recommended. But often the animal handlers become dissatisfied with working conditions when they cannot see the outdoors, even if it is only through sealed window glass.

Animal rooms should be capable of an adjustable temperature range between 65 and 84°F and a relative humidity range between 30 and 50%. In animal rooms containing many closely spaced cage racks, uniform ventilation rates and temperatures from top to bottom and side to center are difficult to maintain. Special HVAC arrangements may be needed to assure that every animal will be maintained continuously under the preselected environmental conditions. Because of animal odors, it is not advisable to recirculate air from animal rooms. Therefore, all of the ventilation air should be discharged to the atmosphere after a single pass through the animal quarters. It may be necessary to filter the air prior to discharge to remove animal hair, bedding fragments, and feces. Certain procedures with animals should be conducted in biological safety cabinets, or in a biological safety laboratory (Chapter 12).

18.1.2 Work Activities

The activities performed in a small-animal laboratory facility include ordinary good animal care of a maintenance and preventive nature; animal experimentation involving administration of drugs, chemicals, and biological agents by inhalation, ingestion, injection, skin application, and surgical procedures; routine pathology preparations and examinations; and record keeping of a highly detailed nature. Much of the daily laboratory routine is occupied with animal care duties such as food preparation, changing animal bedding, washing cages and room surfaces, inspecting animals for illness, registering deaths, and disposing of animal carcasses and other wastes. Direct experimentation with animals may take place inside the animal laboratory when adequate facilities for such work have been provided. Otherwise, the animals may be moved to adjacent laboratories for scientific and surgical procedures, and returned to the animal laboratory daily for housing and routine care. When animals are going to be exposed to hazardous substances or subjected to dangerous procedures, facilities for conducting such work should be made an integral part of the animal laboratory to avoid unnecessary spread of toxic substances and exposure of personnel who are not directly involved.

Only small amounts of drugs, laboratory chemicals, and bottled gases are needed in most modest small-animal laboratories but substantial quantities of sanitizing and disinfecting materials may be stored for routine use. The nature of the latter materials should be checked carefully to avoid inadvertantly introducing toxic or dangerous substances into the animal laboratory.

18.1.3 Equipment and Materials Used

As a minimum, animal laboratories contain cage racks for housing the animals, a cage washing and sterilizing machine, a steam sterilizer for surgical supplies and other equipment, food preparation machines such as scales, dry feed mixers and vegetable slicers, refrigerators, microscopes, surgical and autopsy tables, sinks, a deep freeze cabinet for storage of dead animals, and work benches for holding a variety of scientific instruments used to measure animal responses and examine animal tissues. There may also be items usually found in biological safety laboratories such as high-speed blenders, sonicators, and lyophlizers. For handling these devices, reference should be made to Chapter 12.

18.1.4 Exclusions

Modest small-animal laboratories are not suitable for work with highly infectious biological agents or with more than very small quantities of toxic chemicals, or for housing dangerous animals. Each of these conditions calls for a highly specialized facility that is not covered in this chapter.

For protection of the animals housed in the laboratory and for safeguarding the integrity of the experiments underway, the introduction into and storage of volatile, flammable, and explosive materials in the animal laboratory should be strictly prohibited.

18.2 LABORATORY LAYOUT

18.2.1 Introduction

The minimum-sized small-animal laboratory is a suite of several rooms isolated from the rest of the building by doors and out-of-traffic paths. The animal laboratory should be further isolated by maintaining it under negative air pressure relative to connecting corridors and adjacent rooms to prevent the spread of animal-generated odors outside the animal quarters. The recommendations contained in Sections 1.2 and 2.2 are generally applicable to animal research laboratories except as supplemented or modified in the following paragraphs.

A special feature of an animal laboratory is provision of designated "clean" and "dirty" areas and development of a circulation pattern that prevents passage of personnel and equipment from the dirty to the clean side without first passing through a sanitation station. The objective is to avoid introducing infection to the animal colony. A similar purpose is served by providing isolated quarantine and sick bay areas.

Because of the need to bring cage racks periodically to a central cage-cleaning facility, certain of the corridors in the animal laboratory must be unusually wide to accommodate such traffic. A typical layout for a minimum sized small-animal laboratory using a one-corridor system is shown in Figure 18-1. It may be noted from Figure 18-1 that the receiving and quarantine rooms can be reached from the outside, without entering any other part of the laboratory. The central corridor serves as both a clean and dirty corridor. A better arrangement, the two-corridor system, which utilizes distinct clean and dirty corridors, is shown schematically in Figure 18-2. The important features of a two-corridor system are (1) corridors confined to one-way movement of personnel, animals, and equipment—always from "clean" to "dirty," and (2) two doors to each animal room—one for entry, one for egress.

In small-animal laboratories, the operational rigidity imposed by the two-corridor system can become an obstacle to efficiency and speed. Nevertheless, long-term experiments, such as bioassays for carcinogenicity, that call for the animals to live out a time approaching a normal life span, require the ultimate in animal protection, and the two-corridor system is designed for such purposes. Because of the design rigidities im-

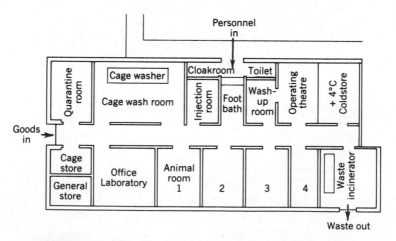

FIGURE 18-1. Plan of a conventional animal house (Inglis, 1980).

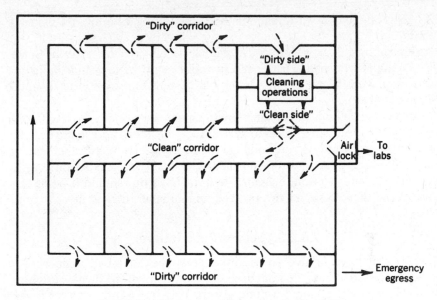

FIGURE 18-2. Two-corridor system (Inglis, 1980).

posed by the two-corridor system of animal laboratory construction, identification of a need for such a facility must be made early in the design process.

18.2.2 Floors, Walls, Ceilings

Animal laboratories require impervious surfaces and structural joints that are vermin-proof and easily cleaned and decontaminated. Walls and floors should be monolithic and made of washable and chemically resistant plastic, baked enamel, epoxy, or polyester coatings. The monolithic floor covering should be carried up 8 in. of the wall to prevent accumulations of dirt and wastes in the corners. Corridors subject to heavy traffic from transportation of cage racks and hand trucks handling feed and wastes must be constructed of materials resistant to wear and frequent washing with detergents and disinfectants. Smooth, hard-surfaced concrete and neoprene terrazzo are often recommended for floors, but no ideal floor construction method or material can be identified as totally trouble-free; frequent maintenance and repairs are required.

Buffer strips on walls in corridors, animal holding rooms, and so on, that will prevent cage racks and hand carts from colliding with the walls, and thereby gouging the surface and rupturing the monolithic coatings, will go a long way toward maintaining a sanitary and vermin-free animal laboratory.

Floor drains are not essential in animal rooms housing rodents. Some veterinarians who are experienced in rodent care believe that moisture associated with the use of floor drains for cleaning purposes is detrimental to animal health.

Suspended ceilings hide from view the many services to the laboratory and produce a pleasing, finished appearance but they represent a serious impediment to pest control. Solid ceilings are preferred, with the heating and ventilating ducts, water and electrical services, and so on, easily accessible for inspection, cleaning, and disinfection (should it be necessary to run these services through the animal laboratory at all).

18.2.3 Access Restrictions

Access should be limited to essential personnel to avoid unnecessary exposure of the animals to infections and contamination. Illegal activities of certain so-called animal welfare groups may make it prudent to maintain the animal laboratory behind locked doors.

Care must be taken to prevent the entry of wild rodents that will be attracted by the availability of food, because they bring diseases usually absent from carefully managed laboratory colonies. This is done by keeping laboratory access doors closed when not being used for passage of personnel, supplies, or equipment, and making certain that the crack under the door will not permit passage of even a small mouse.

18.3 HEATING, VENTILATING, AND AIR CONDITIONING

18.3.1 Introduction

Animal laboratories require rigid control of temperature, humidity, and air movement in animal rooms at all times to provide optimal conditions for the health and growth of the species housed therein. In addition to a need for better than usual HVAC control systems, alarm systems are essential to alert responsible personnel to a system failure long before conditions deteriorate to a level that affects the animals adversely. The importance of alarm systems to signal HVAC dysfunction will be in direct proportion to the duration of the animal experiments that will be conducted. When animal stress from unfavorable environmental conditions must be avoided at all costs, standby HVAC equipment will be needed.

Negative pressure differentials between animal rooms and the remainder of the building housing the laboratory must be maintained to avoid spreading unpleasant animal odors to areas outside the animal laboratory. Within the animal laboratory, pressure differentials should be maintained that always direct airflow from clean areas to dirty areas, never the reverse.

As a rule, air should not be recirculated from animal rooms; it should be discharged to the outdoors, usually after filtration.

18.3.2 Criteria

The HVAC requirements for animal rooms are contained in publications by the National Academy of Sciences, the National Institutes of Health, and by similar agencies in other countries. For rodents, the conditions shown in Table 18-1 are recommended (NAS, 1976). The number of air changes per hour for animal rooms is determined in part by the total animal heat contribution to the environment. Heat gain values for commonly used rodents are shown in Table 18-2. Generally, 10–15 air changes per hour will be needed for animal care rooms containing rodents. If the animal rooms are densely populated, as high a ventilation rate as 20–25 air changes per hour may be required. Such high ventilation rates are undesirable because of the difficulty of providing a draft-free air supply. Room ventilation requirements for commonly used rodents are shown in Table 18-3.

Air conditioning of animal quarters is a complex consideration involving temperature, relative humidity, and air changes, each of which is

TABLE 18-1
Recommended Temperature and Relative Humidity
for Common Rodents

	Temperature		Relative
Rodent	°C	°F	Humidity (%)
Mouse	20–24	68–75	50–60
Hamster	20–24	68–75	40–55
Rat	18–24	65–75	45–55
Guinea pig	18–24	65–75	45–55

Source. ILAR News, Vol. XIX, No. 4, Summer 1976, National Academy of Sciences, Long-Term Holding of Laboratory Rodents, Institute of Laboratory Resources.

TABLE 18-2
Heat Gain Values for Common Rodents

Rodent	Wt (g)	Heat Gain (kcal/h)	Heat Gain[a] (Btu/h)
Mouse	21	0.403	1.599
Hamster	118	1.470	5.833
Rat	250	2.581	10.242
Guinea pig	350	3.322	13.182

[a] Calculated.

Source. *ILAR News,* Vol. XIX, No. 4, Summer 1976, National Academy of Sciences, Long-Term Holding of Laboratory Rodents, Institute of Laboratory Resources.

known to influence the physiological well-being of the animals irrespective of whether these effects are exerted independently or in combination. Control of environmental conditions in animal facilities may affect not only the usefulness of the animals as a research subject but also the quality of the data obtained from the research efforts (NAS, 1976).

18.3.3 Filtration

Exhaust air from animal housing and treatment rooms should be filtered before discharge to remove hair, bedding, feces, and so forth. An NBS 85% efficiency filter will be adequate for this service unless the animals are harboring human pathogens. In that event, secondary filtration through HEPA filters will be required in addition (see Chapter 12). If the

TABLE 18-3
Room Ventilation Recommendations for Common Rodents

Rodent	Wt (g)	Ventilation (m³/h/animal)	(ft³/min/animal)
Mouse	21	0.25	0.147
Hamster	118	0.69	0.406
Rat	250	1.38	0.815
Guinea pig	350	1.97	1.15

Source. *ILAR News,* Vol. XIX, No. 4, Summer 1976, National Academy of Sciences, Long-Term Holding of Laboratory Rodents, Institute of Laboratory Resources.

animals are to be maintained in a germ-free environment, HEPA filtration of all supply air will be required. When using trace quantities of especially toxic volatile chemicals in conjunction with biological experiments, it may be prudent to add an efficient adsorber (generally activated carbon) as an additional effluent air cleaning stage.

18.3.4 Controls

Pressure control can be maintained within an animal laboratory by providing a constant ratio of supply to return-plus-exhaust air with the aid of differential pressure controllers, modulating dampers, fan inlet vanes, or a combination of all of these. Airflow variations should be minimized as an aid in providing control over room pressure. Continuous operation of hoods, cabinets, and local exhaust systems is an important aid in maintaining reliable pressure control at all times. Air locks to adjacent areas may be required. The specifications for air locks should be the same as those required by the General Services Administration (GSA, 1979).

Temperature and humidity control on the other hand may require a constant volume reheat system. All aspects of controls must be considered before deciding on a final control strategy.

Certain built-in safety aspects could be considered. For example, use of N.O. (normal open) control valves on reheat coils should be avoided, otherwise loss of control signal may overheat the animal space.

18.3.5 Alarms

Alarms should be provided for the HVAC systems as outlined in Sections 1.4.5.2 and 1.4.5.3. An additional alarm should be provided to notify personnel when filters are becoming dust-loaded to a critical point. The information on air cleaning system monitoring instruments and alarms contained in Sections 8.3.2.4 for the clean room laboratory also applies to animal laboratories.

18.4 LOSS PREVENTION, INDUSTRIAL HYGIENE, AND PERSONAL SAFETY

The recommendations for loss prevention, industrial hygiene, and personal safety contained in Sections 1.4 and 2.4 apply generally to animal laboratories.

18.4.1 Personal Protective Equipment

Work around animals sometimes results in sensitization to animal dander and other animal products. Should this occur, disposable dust respirators, approved by NIOSH, will generally relieve the symptoms. They will prevent initiation of allergy if worn as a prophylactic measure. Air supply respirators may need to be used in more severe cases of animal allergies.

Personal cleanliness is essential to avoid infection and it is recommended that animal laboratory personnel, and especially those who care for the animals, wear laboratory coats or coveralls and use gloves. These garments should not be taken out of the laboratory. Therefore, provision for storage of clean garments and safe disposal of those that become soiled must be delineated at the laboratory design stage. Also, convenient face- and handwashing facilities are needed inside the animal laboratory for maintaining personal hygiene. Shower and clean clothes locker rooms for animal care personnel are highly recommended.

18.4.2 Decontamination

It is sometimes necessary to decontaminate entire animal laboratories when stubborn animal diseases take over. Decontamination is normally conducted by releasing formaldehyde vapors into a space that has been thoroughly isolated from surrounding areas to prevent escape of formaldehyde vapors to occupied areas. National Sanitation Foundation Standard 49 contains a "Recommended Microbiological Decontamination Procedure" (NSF, 1983).

18.5 SPECIAL REQUIREMENTS

18.5.1 Illumination

Certain animal species follow a diurnal pattern of nighttime activity and daytime rest. For these, provisions must be made for reversing the usual illumination sequence.

Illumination levels recommended by the National Institutes of Health for rodents are generally 60–80 ft-c.

Fluorescent lighting is recommended because it generates less heat than incandescent lamps of equivalent wattage and hence has less effect on HVAC requirements.

PART
III

ADMINISTRATIVE PROCEDURES

19

Bidding Procedures

19.1 Introduction

To ensure that all of the safety and health considerations will be incorporated into the completed laboratory, building, or renovated area, it is of utmost importance that the criteria delineated in this manual be incorporated into the final design specifications put out for bid. It is equally important that correct bidding procedures be established and closely followed to assure construction of the facilities as designed. To that end, a set of bidding documents must be prepared. The following guidelines can be used for their preparation. They should be carefully reviewed by a qualified health and safety professional for appropriate additions or deletions.

19.2 BIDDING DOCUMENTS

Bidding documents can be divided into three major sections: (1) contract forms, (2) general conditions, and (3) technical specifications. Contract forms and general conditions address the legal and administrative requirements of the project and vary according to the size of the project and its location. Requirements for adhering to all applicable federal, state, and city health and safety regulations during construction should be included, and all unusual hazards, including those likely to be present as a result of

TABLE 19-1
Topical Outline for Preparation of Technical Specifications with Notes Identifying Areas of Health and Safety Concern

Division 1—General requirements
  Section 1A—Special conditions[a]
  Section 1B—List of contract drawings
Division 2—Site work
  Section 2A—Preparation and demolition[b]
Division 3—Concrete
Division 4—Masonry
  Section 4A—Masonry
Division 5—Metals: Structural and Miscellaneous
  Section 5A—Miscellaneous metal
Division 6—Wood and plastics
  Section 6A—Carpentry
  Section 6B—Laboratory equipment and millwork
Division 7—Moisture protection
  Section 7A—Roofing and flashing
  Section 7B—Caulking
Division 8—Doors, windows, and glass
  Section 8A—Metal frames
  Section 8B—Wood doors
  Section 8C—Finish hardware
  Section 8D—Metal windows
  Section 8E—Glass and glazing
Division 9—Finishes
  Section 9A—Lathing and plastering
  Section 9B—Seamless flooring[c]
  Section 9C—Painting[d]
Division 10—Specialties
  Section 10A—Miscellaneous specialties
Division 11—Equipment[e]
Division 12—Furnishings
Division 13—Special construction
  Section 13A—Temperature-controlled room
Division 14—Conveying systems
Division 15—Mechanical
  Section 15A—Special requirements for mechanical work
  Section 15B—Plumbing work
  Section 15C—Piping
  Section 15D—Waste system
  Section 15E—Domestic water system
  Section 15F—Plumbing fixtures
  Section 15G—Special and miscellaneous
  Section 15H—Insulation

TABLE 19-1 (*Continued*)

Section 15I—HVAC Work
Section 15J—Piping
Section 15K—Heating
Section 15L—Sheet metal
Section 15M—Insulation
Section 15N—Temperature controls
Section 15O—HVAC systems adjustment
Division 16—Electrical
Section 16A—Special requirements for electrical work
Section 16B—Electrical work
Section 16C—General wiring and equipment
Section 16D—Primary work and materials
Section 16E—Secondary distribution
Section 16F—Grounding
Section 16G—Lighting
Section 16H—Heating
Section 16I—Special systems
Section 16J—Miscellaneous electrical work
Division 17—Final acceptance[f]

Extracted from Uniform Construction Index, Data Filing Format. Master Spec 2, AIA Service Corp.

[a] Any condition that may present a hazard to construction workers should be noted in this section with the procedures to be taken to eliminate the hazard. This section will occur most frequently for renovations where normal laboratory functions will be continuing in nearby areas during renovation. Procedures must be established to ensure that laboratory workers are not subject to a potential risk as a result of the construction, and that construction workers are not exposed to chemical, biological, or radiation hazards as a result of ongoing laboratory functions.

[b] Particular attention must be directed to demolition of older buildings or areas of buildings. The presence of asbestos-containing material must be noted and procedures for handling it safely presented to the contractor(s). All areas (fume hoods, ducts, bench tops) that may be contaminated with hazardous materials should be decontaminated prior to removal or relocation. If this is to be part of the contract, safe procedures should be outlined and an on-site resource person identified.

[c] If epoxy floor covering is being installed in a renovated area while personnel are in adjacent areas of the building, provisions should be made for adequately venting the area.

[d] Note c also applies to areas using oil-based paints or any other construction material that can generate toxic or malodorous fumes.

[e] Special consideration must be given to specifications for laboratory equipment (e.g., fume hoods, biological safety cabinets, refrigerators, centrifuges).

[f] Specifications for performance criteria and final acceptance outlined in Chapter 20 should be incorporated into the technical specifications.

construction activites, should be identified here. These unusual hazards may include the presence of asbestos-containing materials during renovation or demolition. The technical specifications may include many of the statements contained in this manual along with non-safety-related items. An example list is given in Table 19-1. A careful review of all sections should be made by health and safety professionals to identify all health and safety issues. Those sections requiring special on-site inspection have been indicated in the notes that follow Table 19-1.

20

Performance and Final Acceptance Criteria

20.1 GUIDING CONCEPTS

20.1.1 Introduction

The intent of this chapter is not to describe the entire building construction acceptance and testing procedure, but to concentrate on the building systems that pertain to health and safety. It is assumed, for example, that all structural, foundation, and soil testing necessary for a structurally sound building has already been done. It is critical that the health and safety systems in the laboratory building or renovated space be carefully inspected periodically during construction, because it is easier and more economical to correct defects during this stage than to wait until the final acceptance inspection has been made. Such items as proper welding techniques, use of correct construction materials, and precise location of safety equipment storage areas should be observed. Close cooperation between designer, contractor, tradesmen, and final user is necessary for best results. For certain systems (e.g., fire suppression) it is not possible for a useful inspection to take place during construction because, for these, only a complete system-wide test after completion will be meaningful. Nevertheless, it is appropriate to keep abreast of preparations of the assigned space and the installation of auxilliary services that will be required for correct operation and ease of maintenance. For large projects, an on-site engineer responsible to the building owner should be present continuously during the entire construction period to perform the ongoing construction inspections in a timely manner. It is recommended that this person be appointed in addition to the normal clerk of works. Some suggested test procedures follow. They should be included in the bidding documents.

20.1.2 Regular Testing Prior to Occupancy

To ensure that emergency systems will perform satisfactorily when needed, it is essential that frequent testing be conducted even during the interval before the building is occupied. This is commonly carried out once a week and should include the emergency electrical systems, the fire alarm systems, and all other emergency alarm systems when in service.

20.2 DESIGN, CONSTRUCTION, AND PREOCCUPANCY CHECKLISTS

The list presented below can serve as an aid in focusing on major safety and health items that need to be addressed during the various stages of laboratory construction. This is by no means intended as a complete list but it will serve as an initial checklist to be added to depending upon the special needs of the particular project.

20.2.1 Planning and Design Safety Review Checklist

- Provide adequate chase space for currently specified exhaust ductwork, compressed gas piping, and so on, and allow at least 25% additional space for future additions.
- Avoid horizontal runs of ductwork in all exhaust systems susceptible to internal condensation. Perchloric acid hoods are a special hazard when condensate can collect inside the ducts. Do not combine exhaust ducts from perchloric acid hoods with other exhaust air systems.
- Select locations for portable fire extinguishers and identify the type of units at each location.
- Plan laboratory bench layout for rapid emergency egress.
- Select and identify locations for eye protection dispensers at entrance to eye-hazard areas (e.g., at laboratory doors).
- Select doors with glass panels so that oncoming traffic can be seen.
- Avoid copper piping for acetylene gas lines.
- Document the safety and health practices that contractors and their subcontractors are expected to comply with while working on the site.

- Select and designate areas requiring emergency lighting capabilities. Restrooms and some laboratory areas are frequently overlooked although they are included in NFPA codes.
- Plan fire escape routes.
- Select and designate locations for fire alarms
- Plan height of storage shelves so that sprinkler systems are not compromised.
- Plan for installation of waste containers that segregate combustibles, noncombustibles, chemicals, hazardous waste, broken glass, and trash.
- Plan special facilities for use and storage of cryogens (e.g., liquid nitrogen) when these products are needed.
- Identify the systems that will require backup emergency services (e.g., electricity, cooling water, compressed gases/air) and make certain these facilities are included in the plan.
- Identify and record where explosion-proof or laboratory-safe refrigerators are required.
- Select and designate locations for safety closets and first aid facilities.
- Identify outlets for potable water and special laboratory water.
- Plan convenient locations for and identify shutoff valves for all piped utilities, including water services.
- Plan for completing and issuing a site safety handbook when occupancy begins.
- Plan facilities for receipt, storage, handling, and disposal of radioactive materials and initiate paperwork to obtain the required license.
- Plan for purchase and storage of necessary personal protective devices such as self-contained breathing apparatus and protective clothing. Designate locations of emergency equipment closets, lab safety equipment drawers, and first aid rooms.
- Choose floor materials that are resistant to slipping and resistant to spills of the chemicals and petroleum products that are likely to be used in large amounts.
- Determine whether exhaust air from some fume hoods will have to be cleaned before being emitted to the environment. When air cleaning is needed, the type, size, and location of the equipment, and the utilities required (water, sewage, electricity, etc.) should be designated in the plan.
- Select the location of air supply intakes away from the influence of exhaust stacks, parking lots, and other sources of contaminated emissions, including those from adjacent buildings.

- Plan for segregated storage of small amounts of oxidizing and combustible chemicals in laboratories and for major stores of chemicals and gases in segregated and specially constructed gas and chemical storage sheds. Review compressed gas storage design for compliance with legal, insurance company, and fire protection association standards.
- Plan to install warning signs in conspicuous locations to identify dangerous areas.
- Plan for and identify locations of an adequate number of emergency exit doors.
- Plan for and identify locations for ground fault interruptors wherever electrical shock hazards may exist.
- Select an incinerator, if required, that will comply with local emission codes and will have 150% of capacity based on currently anticipated needs. Locate a site that is isolated from public access and has ample secure storage space for unburned waste and ashes. Plan for fireproof and vermin-proof construction and plan for a high-temperature flue that is isolated from all fresh air intakes.

20.2.2 Construction Safety Review Checklist

- Make certain that all solvent storage cabinets are electrically grounded by inspection and test.
- Make certain that all electrical outlets in laboratories are labeled and that corresponding labels are provided at each panel box prior to occupancy.
- Make certain that all fans, ducts, air cleaning devices, and the hoods they serve are labeled with coded tags for easy identification.
- Make certain that all specified flow indicators have been installed in the correct location in every local exhaust system.
- Make certain that the bottled compressed air backup system for equipment that would be damaged by a loss of compressed air is installed and connected to all designated delivery points.
- Clean out all compressed gas lines with a nonflammable solvent followed by compressed nitrogen drying before hookup to compressed gas sources. Cleanliness testing is recommended by microscopic examination of white membrane filters through which 1 m^3 of compressed gas has been passed after passage through the longest branch of the piping systems.
- Make certain that all piping systems have been clearly marked in a readily visible area according to an acceptable standard coding system for easy identification.

20.2.3 Preoccupancy Safety Review Checklist

- Test eye fountains and safety showers.
- Test audibility of fire evacuation alarm system.
- Verify direction of door swing with respect to emergency egress routes.
- Check cup sinks for strainers.
- Check that all equipment items, such as sinks, compressors, cabinets, and shelves, are bolted down.
- Review all ventilation system balancing records and make certain that all systems are certified to be in conformance with all applicable plans and specifications.

20.3 HEATING, VENTILATING, AND AIR CONDITIONING

20.3.1 Air Balancing

In certain types of laboratories where pressure relationships are especially critical (e.g., biosafety laboratories), the tightness of the room enclosure is very important. All penetrations into the room (pipes, electrical conduits, ducts) must be well sealed. A leakage test, using sulfur hexafluoride tracer gas, may be appropriate when total containment is required.

For ventilation systems to perform as designed, it is necessary that the supply and exhaust systems be balanced after installation is completed. Balancing will necessitate fan tests that include measurements of static pressure, fan and motor rpm, air volume rate, temperature rise, current draw (to determine brake horsepower), and so forth. Adjustments of sheaves, dampers, and so on, will also be necessary to distribute air in accordance with the HVAC specifications. All balancing should be conducted in accordance with the standards of SMACNA and a written report submitted for the design engineer's review and approval. A sample test report is shown in Table 20-1.

The contract mechanism for obtaining a balancing contractor is a special concern. One approach is to insist that the testing and balancing contractor work for the owner, and thereby have autonomy and the ability to report freely to the owner or his representative any problems noticed with regard to the work of the mechanical contractor. The other approach is to have the testing and balancing contractor work directly for the mechanical contractor on the assumption that this arrangement leads to closer coordination and greater effectiveness. Our recommendation is

TABLE 20-1
Air Moving Equipment Test Sheet

AIR MOVING EQUIPTMENT TEST SHEET

Date _____

Project _____ Project Number : _____

	SPECIFIED	ACTUAL	SPECIFIED	ACTUAL	SPECIFIED	ACTUAL	SPECIFIED	ACTUAL
SYSTEM NO								
LOCATION								
MANUFACTURER								
MODEL NO								
SERIAL NO								
OPERATING CONDITIONS	SPECIFIED	ACTUAL	SPECIFIED	ACTUAL	SPECIFIED	ACTUAL	SPECIFIED	ACTUAL
TOTAL C F M								
RETURN C F M								
O S A C F M								
EXHAUST C F M								
TOTAL STATIC								
SUCTION STATIC								
DISCHARGE STATIC								
EXTERNAL STATIC								
B H P								
MOTOR MANUFACTURER								
SIZE (HP)								
VOLTAGE								
RPM MOTOR								
SAFETY FACTOR	RATED	RUNNING	RATED	RUNNING	RATED	RUNNING	RATED	RUNNING
AMPERAGE								
RPM FAN								
SHEAVE POSITION								

for the testing and balancing contractor to work directly for the owner and that a sum of money be designated for this service function right from the start. We do not recommend that any contractors be permitted to monitor their own work.

20.3.2 Fume Hoods

The face velocity of all fume hoods should be checked to ensure that the airflow rate is in conformity with design requirements. Appropriate label-

TABLE 20-2
Laboratory Fume Hood Inspection Form

BUILDING   DEPARTMENT   ROOM NUMBER   DATE

HOOD NUMBER   PERSON IN CHARGE   LOCATION OF HOOD IN ROOM

USE OF HOOD: HOW OPERATED

RADIOACTIVE MATERIALS   MANUFACTURER_____

PERCHLORIC ACID   TYPE OF HOOD_____

GENERAL CHEMISTRY   SASH: Vertical

HIGH HAZARD CHEMISTRY   Horizontal

SPECIAL PURPOSE

RECOMMENDED SASH HEIGHT

VELOCITY FPM_____

HEIGHT_____

DATE_____

SMOKE TEST_____   HOOD MEASUREMENTS

BOTH SASHES OPEN			ADJACENT SASH OPEN	

AV. VEL._____   AV. VEL._____   AV. VEL._____

EXHAUST FOR CABINET   AIRFOIL_____   AV. VEL._____

      HIGH   LOW   _____   AV. VEL._____

LEFT   ____   ____

RIGHT   ____   ____   SMOKE TEST_____

SMOKE
TEST   ____   ____   COMMENTS

                                  DATE OF LAST SURVEY_____

MICROSWITCH_____   HEIGHT OF SASH_____

ALARM_____

MAGNEHELIC   LOW_____ HIGH_____" H_2O

ing is required on each fume hood to indicate correct operation before it is released for service. If the fume hood is an auxilliary air type, it is necessary to test for correct operation of the makeup air system by smoke trails. Fume hood field performance tests are discussed in Appendix Vh. If the fume hood has an integral flammable liquid storage cabinet, selection of the correct fire rating should be verified. Correct installation and operation of all piped-in gases and electrical fixtures associated with the fume hood should be verified. In hoods equipped with HEPA filters, the integrity of filters and filter mounting should be verified by in-place testing using the standardized techniques recommended for biological safety cabinets in National Sanitation Foundation Standard 49 (NSF 49, 1983) or the techniques recommended for nuclear applications in ANSI/ASME N510 (ANSI, 1980). An example report is included in Table 20-2. When two-speed, variable-speed, or parallel fan types of arrangements are used for exhausting a fume hood, it is necessary that proper operation be verified in each separate mode.

20.3.3 Duct Work Testing

All exhaust duct work should be tested to ensure that excessive leakage does not occur. This is most important when the exhaust ducts are under positive pressure and may leak contaminants into occupied areas. ANSI/ASME Standard N510-1980 gives approved procedures for testing the leak-tightness of exhaust ducts and plenums.

20.4 LOSS PREVENTION, INDUSTRIAL HYGIENE, AND PERSONAL SAFETY

20.4.1 Fire/Smoke Alarms

The complete fire alarm system in the building should be checked for correct operation. Each device should be checked individually and as a part of the system by simulation of alarm conditions. Procedures for testing fire alarm systems are described in NFPA 72 (NFPA, 1985).

20.4.2 Other Alarm Systems

The correct operation of all other alarm systems should be verified. Alarms are frequently used to signal unbalanced airflow, improper operation of mechanical equipment, and so on. Each device should be checked individually and as part of the entire system by simulation of alarm conditions to ensure correct operation even at remote station monitors.

20.4.3 Emergency Electrical System

The emergency electrical generator and associated electrical systems should be started and tested under appropriate load conditions and the engine operated for at least 3 h under 100% overload to ensure that the system will operate as specified when called into service. All transfer switches and ancillary devices should be tested individually and as part of the system and not accepted until found satisfactory.

20.4.4 Eye Wash Facilities

The water flow rate of tempered and nontempered eyewash stations should be verified and recorded, and the angle and height-of-rise of the streams should be documented. Tempered eyewash stations should be checked to ensure water temperature is $70 \pm 5°F$.

20.4.5 Emergency Showers

Flow rate should be measured and recorded. Minimum acceptable flow rate is 30 gal/min. Temperature of tempered showers should be between 70 and 90°F.

21

Energy Conservation

21.1 INTRODUCTION

In this chapter methods for achieving maximum energy conservation are discussed. This chapter should be reviewed during the planning stages of a new laboratory building or during renovation planning for an older building. The selection of one or another of the methods that will be presented will depend on a number of factors. For renovation projects it will include age and geographical location of the building, spatial organization of laboratories and building, size of renovation project, available capital and desired payback period, work schedules, and projected building use. The discussion of the major areas of energy conservation will be divided into five categories: (1) exhaust ventilation for in-laboratory contamination control by the use of chemical fume hoods, biological safety cabinets, and local exhaust points; (2) general laboratory ventilation; (3) lighting; (4) thermal insulation; and (5) humidity control.

Laboratory buildings tend to be energy intensive. Good engineering practice should allow shutdown of laboratory systems when not in use. Careful control of heating systems in laboratory buildings results in less overheating and also saves operating costs.

21.2 ECONOMICS OF EXHAUST VENTILATION FOR CONTAMINATION CONTROL

A major operating expense for most laboratory buildings is associated with the operation of chemical fume hoods. Therefore, major energy savings can be made by the selection of fume hoods that minimize loss of conditioned air. In addition, operational considerations, such as exhaust air quantity and period of operation, play an important part in determining the energy cost of essential laboratory services.

The laboratory chemical fume hood is the major piece of safety equipment available to laboratory personnel who must, from time to time, work with hazardous chemicals and/or hazardous biological agents. It is estimated that at 1985 energy costs, a laboratory fume hood costs between $2000 and $4000 per year based on 24-h/day, 7-day/week operation. The range is a reflection of the variation in the size and design of the fume hoods that are installed. The average cost is approximately $3.60/cfm based on the January 1981 cost of electricity and fuel in the Boston area. (DiBerardinis, 1983) When there are 700 hoods at a facility, using an average cost of $3000 per hood, the operating cost is over $2,000,000 per year. It will, therefore, be clear that substantial energy savings can be realized by installing the most energy-efficient hoods, and equivalent savings can be realized by installing new hoods in older laboratory buildings.

21.2.1 Alternative Energy-Saving Methods

Looking at laboratory chemical fume hoods from an energy conservation point of view, there are four basic alternatives to be evaluated:

1. Reduce operating time.
2. Limit the air quantity exhausted from each hood.
3. Use auxiliary air hoods.
4. Use heat-recovery systems.

We will evaluate the advantages and disadvantages of each option from an economic and safety viewpoint.

21.2.1.1 Reduce Operating Time

Many laboratory chemical fume hoods are presently operated 24 hours a day, 7 days a week. The main reasons for this are

1. Laboratory personnel work in the laboratories all times of the day and night as well as on weekends and holidays.

2. Volatile hazardous chemicals are stored inside hoods when they are not being used. In fact, a need to confine and vent these packages is cited as an important reason for operating hoods in a continuous mode.

3. Some reactions and preparations must be continued in the hood, uninterrupted, for 24 h or longer.

The philosophy behind reduced hood operating time is that when there is no need for a hood to operate (that is, when no one is working actively in a hood, or there is no long-term reaction or preparation taking place), there is no need for that hood to be exhausting the amount of air that it normally does. To reduce operating time, when the hood is no longer needed it is essential either to shut off the fume hood entirely or to lower the quantity of exhaust air. To accomplish this, it is necessary to establish routine working hours when hoods and exhaust points will be fully operational and to establish a procedure whereby legitimate needs during off-hours can be accommodated. It is also necessary to arrange for alternative safe storage for volatile toxic and other hazardous chemicals, including compressed gases.

It is estimated that laboratory fume hoods can be shut off at least 50% of the time without serious interference with research and teaching activities. The advantage is that it can provide a large energy savings. There are also disadvantages that need to be evaluated carefully and the plan adapted to individual circumstances; for example:

1. Hood use will be restricted. This can be overcome by providing local control, that is, by giving each user the option of turning the hood on whenever it is needed during off periods. This is best handled by a central building service group because individual control at the hood usually results in the hood never being turned off.

2. The need for volatile chemical storage can be satisfied by providing alternative safe storage space for the hood user. This can be achieved in several ways: One is to use flammable-liquid storage cabinets that meet NFPA, FM, and OSHA requirements. A second is to provide a separate storage area for nonflammable hazardous liquids that can take the form of (a) a separate, exhausted air storage cabinet, (b) storage under a laboratory fume hood provided with a separate exhaust connection that operates continuously, or (c) storage in a specially designed, passive chemical storage box, developed at the Harvard School of Public Health (DiBerardinis, 1983). An initial capital expense will be required to provide these facilities for each laboratory. In some cases, there may be impediments due to space limitations and/or competing HVAC requirements in ex-

hausting numerous separate cabinets. Although the air volumes needed to exhaust closed cabinets will be modest, that is, a few cubic feet of air per minute, the Harvard storage box requires no ventilation.

3. A troublesome problem associated with reduced hood operating time is maintaining adequate air balance within the laboratory and the building. Many buildings have one supply system for the entire building, or for each large group of laboratories. The difficulty in maintaining correct pressure relationships when hoods and exhaust points are shut down (that is, maintaining laboratories negative with respect to corridors and maintaining hazardous areas more negative than nonhazardous areas) may be difficult to overcome in certain types of buildings and in laboratory wings of multifunctional buildings. This matter has to be evaluated closely for renovations. In most cases, a modulating air supply system that responds to changes in exhaust air demand will be needed. It is possible to resort to a two-step supply air system on the assumption that not more than a few hoods will be operational during off hours and, therefore, the normal building requirement for outside air will provide sufficient supply air for the few exhaust facilities in operation.

The safety problems associated with reduced hood operating time are encompassed in the three points discussed above, the primary one being provision of alternative storage for hazardous materials.

21.2.1.2 Limit the Air Quantity Exhausted from Hoods

The basis for limiting hood air quantity is that under present good practice conditions, the exhaust air requirement is based on the largest possible hood opening. The largest possible opening is usually the length of the work surface of the fume hood times the height of the fume hood opening when the sash is in the fully raised position. But the maximum possible hood opening can be reduced in two ways:

1. Limit the height of the vertical sash opening. If, for example, under normal conditions, the laboratory fume hood sash can be raised 30 in., the exhaust air requirement must be designed for that full opening. If, however, it were arranged that the sash could only be raised 20 in. there would be a one-third saving in the amount of air to be exhausted. This method of reducing hood face opening may present a problem to hood users because it restricts access to the upper part of the hood. To overcome this restriction, it is possible to equip the hood with an alarm system so that when the sash must be raised above 20 in. for short periods of time to allow installation and construction of apparatus within the hood, an alarm, both audible and visual, will be activated to let the hood user know that the hood is not in a safe operating mode. When the sash needs to be above 20 in. for

longer periods, the audible alarm may be turned off, but as soon as the sash is lowered to or below 20 in. the alarms will be reset and become available to respond whenever the hood sash is raised above 20 in. Twenty inches is an arbitrary height for the sash alarm; it can be varied depending on the needs of individuals. In some cases, less height may be needed, whereas in others, the full opening may be required at all times.

2. Limit the air quantity by the use of horizontal sliding sash rather than vertical sliding sash. Usually, four panels of horizontal sliding sash are used. At any given time, only one-half of the full width of a chemical fume hood is open and only the open area needs airflow. About 50% energy savings can be realized by this hood design. A frequently encountered problem with this type of hood sash is nonacceptance among hood users used to working with the more conventional vertical rising sash. A distinct advantage of horizontal sliding sash over restricting the height of vertical sash is that users can still get to every part of the hood, although they have to move the sash horizontally to do so. An added safety benefit of horizontal sliding sash is that it provides a safety shield that users can work behind by placing their arms around it. From a safety point of view, certain designs produce disruptive air currents at the edges of the horizontal sliding sash and this has to be corrected. Unsafe conditions can be created by the ease with which the horizontal sliding sash can be removed from the hood. A method for monitoring and ensuring that the sash remain in place is needed.

3. Substitute local point exhaust systems for conventional hoods. Certain applications do not require a laboratory chemical fume hood. A local point source of exhaust air fits over such devices as gas chromatographs and may be located on laboratory benches where limited quantities of toxic chemicals will be used. The advantages of local exhaust points are that they exhaust far less air than a conventional hood even when the face opening is restricted and the open end can be directed to capture contaminants at the source of generation. Some of the disadvantages are that they are usually designed for specific applications. When different applications are called for, they are sometimes difficult to adapt successfully. In addition, they do not provide the protection that a fume hood does in terms of containment of spills or protection from small explosions or fires.

4. Use multispeed fans to limit air quantity. At high speed, a fan will provide enough exhaust air to give the design face velocity across the entire open face of the hood but when the hood is not in use and the sash is lowered, the fan will go to a lower speed, only one-third to one-tenth full speed, and exhaust just enough air to prevent escape of vapors from the materials or equipment left in the work space. This air volume will be very much less than needed when the hood is in full use. However,

multispeed motors operating down to one-third to one-tenth speed are special and not readily available. The most common multispeed motors are full speed and half speed. In many cases, the use of half speed may be adequate. Referring to the fan laws, a speed reduction of 50% will result not only in a volume reduction of one-half but also in an electrical power saving of one-half raised to the third power, or to one-eighth the power required at full speed. Variable-speed motors are becoming more readily available.

5. Use a very small fan in parallel with the main exhaust fan. During periods of nonuse, the large exhaust fan can be shut off, leaving the small exhaust fan to provide a low-volume continuous exhaust airflow to maintain safe conditions. Care must be taken to prevent short-circuiting with this fan arrangement.

These systems have a common drawback, that is, the possibility of active hood use while it remains in a dormant or nonuse mode. Various alarm systems have been designed and are commercially available to inform the hood operator of potential unsafe conditions. A similar concept is to install a device that locks the sash closed when the hood is not in a safe operating mode.

Note. Duty cycling of laboratory exhaust systems creates very unsafe conditions and should not be employed for energy conservation purposes.

21.2.1.3 Limit the Volume of Conditioned Air Exhausted from Hoods (Auxiliary Air Hoods)

Auxiliary air hoods are laboratory chemical fume hoods that supply a major part of the total exhaust air volume from a special supply air duct located above the work opening of the hood, with the remainder coming from the general room supply air. Up to 70% of the air that the hood exhausts is provided at the face of the hood and the remainder is provided by the building supply air system. The economic advantage of this hood is that in winter time, the auxiliary air coming into the hood above the hood face needs to be tempered only to 60°F and not to the full temperature of the air supplied to the room. The theory behind this is that all the air is going directly into the hood and not mixing with the room air so that it does not need to be fully heated. It does, however, need to be tempered somewhat because some of the auxiliary air flows down across the head of the person standing at the hood. If it were not tempered to the degree noted, the worker would experience a cold draft. The largest energy saving is realized by the user of these hoods during hot weather in the warmest climates, because in the summertime the auxiliary air does not need to be cooled for comfort. In completely air conditioned buildings, considerable energy can be saved by installing these hoods. Another advantage associated with the use of auxiliary air hoods is that in areas where there are a large number of hoods they eliminate the need to provide large amounts of makeup air that would represent an excessively large number of air changes per hour in the laboratory and result in uncomfortably high air velocities sweeping through the laboratory.

There are several undesirable features of auxiliary air hoods. First, they require an additional mechanical system, which can generate its own maintenance problems. Second, the design of this type of hood is critical. In 1984, a number of hoods on the market, whose manufacturers claim they are auxiliary air hoods, provided the auxiliary supply air inside the hood rather than outside the hood and above the work opening. Introducing supply air inside the hood can result in pressurization of the internal volume and result in toxic materials coming out of the open face of the hood. In other unsatisfactory hood designs, the auxiliary air is supplied above the work opening but it is incompletely captured by the hood. Instead, it mixes with the room air and defeats the energy conservation purpose of the hood. In northern climates, net savings from installation of auxiliary air hoods may not be realized because the period of hot weather is too brief each year to return a saving of energy equivalant to the added installation and operating costs. During the long cold winter, the auxiliary air has to be tempered to a temperature almost as high as the general room air supply and worthwhile savings may not be realized. One hood manufacturer claims the difference between the cost of an auxiliary air hood and a conventional fume hood is paid back in less than a year through energy savings. There may be locations and usage patterns where such energy savings can be attained, but it is by no means a general rule. If an auxiliary air hood is selected, one must be certain it can operate as designed. The performance tests outlined in Appendix Vh should be used as a guide for hood selection.

21.2.1.4 Heat-Recovery Systems

Heat-recovery systems employ some type of heat exchanger to extract heat from the exhaust air stream in winter and use the recovered heat to partially warm the incoming air. The reverse cycle is applied in warm climates. The application of this type of system has to be evaluated on a case by case basis because the climate in the area of contemplated use may not lend itself to a useful amount of energy savings with these kinds of systems. It should be kept in mind that although a worthwhile savings may be calculated for the days of extreme temperature excursion, such days are often too few each year to pay back the cost of the installation plus the energy cost associated with operating the heat-exchanger system. The additional pressure drop induced by all these devices leads to increased horsepower requirements that need to be considered in the economic analysis. In addition, maintenance and repair requirements when recovering heat from laboratory exhaust air tend to be higher than normal.

Because laboratories are thought by some to be energy wasteful, some administrative codes mandate heat recovery from laboratory exhaust systems even though small temperature differences make heat recovery inefficient thermodynamically.

Methods available for air-to-air heat recovery are

1. Run-around loops
2. Heat wheels
3. Heat pipe systems
4. Plate heat exchangers
5. Air-to-air heat exchangers (not acceptable due to contamination and other problems)
6. Heat pumps
7. Chemical regenerators

Some of these systems are illustrated in Figure 21-1.

Exhaust air from biomedical laboratories may have special requirements that must be addressed:

1. In some cases the exhaust air must be incinerated to eliminate the danger of biological contamination.
2. The exhaust air from the animal areas must be carefully filtered to eliminate animal hairs, food particles, and so forth, before it enters a heat-recovery device.

Studies have shown that a rotary air-to-air regenerative heat exchanger with either a metallic or desiccant-impregnated matrix is the most efficient device available because of its total heat-transfer capability. Although air-side cross-contamination can be reduced to less than 1% by volume when the device is designed with a purge chamber, the major disadvantage is that the device is subject to exhaust-side condensation during winter operation when serving chemistry laboratories. Water-soluble hydrocarbons and other chemicals will revaporize and enter the makeup airstream producing inlet air contamination. Furthermore, because of the nature of the chemicals used, a reaction may take place that could destroy the metallic

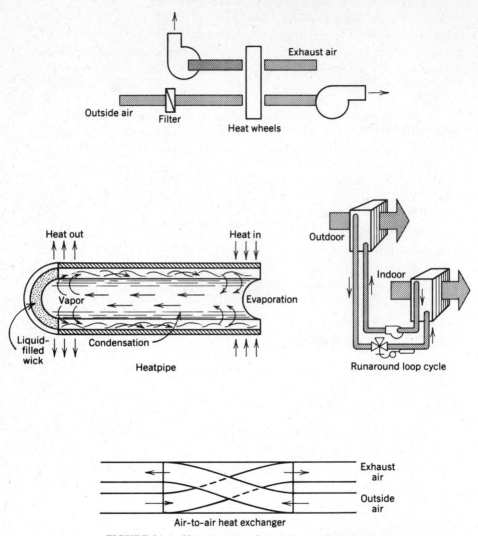

FIGURE 21-1. Heat recovery from exhaust airstreams.

or desiccant material. Experience has indicated that after some period of operation, a substantial amount of leaking occurs between the exhaust and supply sides of this system. Because of these problems, rotary heat exchangers are not recommended.

Only sensible heat-transfer devices are recommended. Stationary plate exchangers provide good separation between makeup and exhaust streams but they are heavy and bulky, and are difficult to build for large installations. In addition, they have only a 40–60% range of efficiency. Coil runaround systems have good potential when makeup and exhaust streams are physically remote. However, the range of effectiveness is also only 40–60%. A nonregenerative heat pipe in a coil configuration can be more effective, ranging from 60–70%, and the airstreams are isolated from each other by a center baffle or separator, preventing cross-contamination. Liquid lithium bromide systems are not used for laboratory air heat recovery because they are unlikely to avoid contamination of the incoming air. With the cost of fuel coming down, it is unclear how long these techniques will remain cost effective.

21.2.2 Conclusions

It can be seen that there are several alternatives that may be applied to realize an energy saving when dealing with laboratory chemical fume hoods. There is no single answer as to which may be the best method. In many cases a combination of methods may be required, particularly in new buildings. In the retrofitting or renovation of older buildings it may be difficult to apply some of the methods that have been reviewed and therefore each situation has to be evaluated as a unique problem. The issues that need to be evaluated to make the best decision include the following:

1. Cost of retrofit for each of the alternatives.
2. Acceptance by hood users of a variety of restrictions that may be placed on hood use.
3. Effect of each alternative on the functioning of the building HVAC systems, on the comfort of the occupants, and on the type of work that must be conducted in the laboratory fume hoods within each building.
4. Effect of each alternative on safety in the laboratory. This factor must be evaluated carefully, including consideration for (a) providing additional storage space, (b) acceptance by laboratory personnel of restricted operating time, and (c) use of different types of laboratory fume hoods.

Any program that seeks to alter the traditional use of laboratory chemical fume hoods must include a detailed education and training program for laboratory personnel that will include information on how a laboratory chemical fume hood operates, what restrictions, if any, are being placed on its use, and how the restrictions may affect their research activities.

21.3 LIGHTING

A reduction in the energy required for lighting can be accomplished in laboratories by using task lighting at desks, laboratory benches, and work stations. The use of energy-efficient fluorescent tubes and ballasts, and multiple switching within a laboratory building will also conserve electrical energy. Lighting should be maintained at the levels outlined in Section 1.5.

21.4 THERMAL INSULATION

This energy-conservation measure is very effective for residences, but is not as useful for laboratory buildings because the heat transferred through the structure is a small fraction of the energy expended for ventilation air conditioning and contamination control. Nevertheless, the energy savings achievable through the use of good building thermal insulation are worthwhile and should be realized.

21.5 HUMIDITY CONTROL

Needs for humidity control should be carefully considered. In summer, for example, to maintain excessively close humidity tolerences or exceptionally low humidity means more intensive use of the air conditioning system because the usual method for reducing humidity in the air is to subcool it to reduce the moisture content and then to reheat it to the required room temperature. Because this process is an enormous energy consumer, it should be avoided wherever possible. To maintain high humidity in winter, moisture must be added to the air, and this is also energy intensive.

Background on HVAC

I.a A DESCRIPTION OF AIR CONDITIONING SYSTEMS

This appendix is intended to provide a general background for the non-HVAC type on the terminology used in describing HVAC systems. It is by no means an exhaustive discussion of HVAC systems. For more detailed information see ASHRAE (ASHRAE, 1983–86). Additional reference sources include McQuiston, 1977, Stoeker, 1958 and Thuman, 1977.

Air conditioning is commonly understood to mean the supply of tempered (heated or cooled) air into a room to offset heat losses or heat gains, but air motion, relative humidity, and air purity controls are also included in a fully implemented system. A conditioned zone is a boundary inside the conditioned space that is controlled by a single control point or thermostat.

To a remarkable degree all warm-blooded animals, including man, are able to maintain a constant internal environment, called homeostasis, while living in a changeable external environment characterized by extremes of temperature and humidity. Man alone, however, is unique in having the ability to regulate the external environment in accordance with his own ideas. Warming the interior of buildings during cold weather has been carried on since time immemorial, but cooling is a twentieth century phenomenon, initiated for factory production of heat- and humidity-sensitive items, adopted on a large scale in the United States for commercial properties during the 1930s, and for residential use after 1950. Although

215

the original objective in cooling commercial and residential properties was to provide a lower indoor temperature during hot weather, it was soon learned that the moisture content of the cooled air and its rate of motion over the body were additional factors in regulating human comfort in air-cooled spaces. This knowledge is expressed in "effective temperature charts" (Appendix II).

Heat energy flows from a region of higher temperature to one of lower temperature, meaning that building interiors lose heat when the weather is cold and gain heat when the weather is hot. The rate at which heat flows from the interior to the exterior or vice versa is variable and depends partly on the conductivity of the structure and partly on the amount of air exchange between inside and outside. Outside, unconditioned air is brought into the structure in two ways: purposefully for ventilation supply air and inadvertantly as infiltration air that enters through open doors, windows, and structural cracks under the influence of an inside pressure that is lower than the pressure outside the structure. Wind action has an important influence on infiltration rates as well as on heat conductivity.

During the past decade, energy conservation measures have been influential in substantially decreasing the heat conductivity of new and old structures by the addition of heat insulation materials, and by reducing building porosity to decrease infiltration. Both measures have been effective in reducing building air conditioning costs and it is unlikely that this trend toward energy efficiency will be reversed in the foreseeable future.

Heat gains by insolation are welcomed during cold weather, but guarded against during hot weather by window and structural shading, reflective roof and wall treatments, and building orientation relative to the sun. Heat gains by insolation are generally neglected when calculating heating requirements but must be carefully accounted for when calculating cooling loads. A similar comment applies to outside air humidity. Heat gains from insolation and the need to reduce the moisture content of outside air entering the air conditioned spaces through mechanical ventilation and infiltration, together often represent a major fraction of the total summer cooling load. Heat sources within the building itself, from lighting, equipment, and personnel, decrease the winter heating load but increase the summer cooling load. In some laboratories, the heat generated internally by equipment, people, and lights may be greater than the summertime heat gains from the sun and by conduction through the building's structure. It is critical, therefore, to inquire closely into the nature and amount of laboratory equipment that emits heat when designing the HVAC facilities that will be included in the building design. In extreme cases, a cooling capability may be required year round, even in cold climates, and this requirement must be provided when the heat load from equipment is unusually large.

The usual design procedure is first to establish extreme operating conditions in order to select equipment capable of satisfactory full-load operation. Next, decisions must be made regarding the type of system required (e.g., multiple zones, individual room controls), and the nature of the equipment that will be selected to deliver the preselected design conditions (e.g., central heating, humidifying, and cooling systems, under-window units to provide final temperature control).

In a *central system* cooling and heating are performed with central chillers and boilers connected to large air handling equipment that serve the whole or a major part of the building. The heating or cooling sources may be remote from the buildings they serve.

In a *modular system,* heating and cooling are done locally, sometimes room by room. Central and noncentral cooling systems used in modern buildings are described below.

I.a.1 Cooling System

The most commonly used cooling system, referred to as mechanical cooling, is a thermodynamic process called direct expansion. A gaseous heat-transfer medium such as Freon or ammonia is compressed to a state of high temperature and pressure by a compressor and then transformed into a cool liquid in a condenser, which may be air or water cooled. The cooled liquid, a refrigerant under high pressure, is allowed to flow to a low-pressure region through an expansion valve where the refrigerant evaporates. The latent heat of evaporation is extracted from the fluid (air or water) being cooled thereby producing the temperature reduction step. The warm evaporated refrigerant gas is drawn next into the suction side of a compressor, and the cycle repeated. This most basic of thermodynamic cycles is the heart of all direct expansion, or DX, refrigeration cooling systems. Some of the variations are described below.

I.a.1.1 Compressors and Prime Movers

There are many types of compressors in use for air cooling systems. Compressors can be hermetic or semihermetic. In hermetic compressors, the electric motor windings are cooled by the refrigerant suction gas. This prevents overheating, and at the same time increases efficiency by heating up the cool suction gas, evaporating any liquid droplets that may be in the stream. A true hermetic compressor is integrally sealed. In smaller sizes it is called a "tin can." A semihermetic compressor is similar to a hermetic one, except that it can be taken apart for maintenance. On larger systems, it is almost mandatory that a semihermetic compressor be used. Another classification is an open compressor where the prime mover operating the compressor is separate from it. Although there is more flexibility in the choice of prime movers, the most frequent choice is an electric motor. Other kinds of prime movers, such as gasoline engines and steam or gas turbines, are also employed.

In large systems, the voltage to the electrical motor may be substantially higher than usual to avoid unusually large wire sizes and motor winding.

Reciprocating Compressors
The most common compressor is a reciprocating one in which the gas is compressed in cylinders by pistons operated by an electric motor.

Centrifugal Compressors
This machine compresses the refrigerant gas by centrifugal action rather than by reciprocal piston action but is otherwise similar to a reciprocal pump in its function. It is used primarily in systems having more than 100 tons of refrigeration capacity because it provides a more efficient ratio of energy consumption.

Screw Compressors
In this type, the refrigerant gas is compressed between two turning helical screws. The system is generally used from medium to high refrigeration tonnage because the energy ratio (i.e., kilowatts of electricity expended per ton of refrigeration produced) is attractive and the system offers a very reliable alternative to reciprocating and centrifugal compressors. The choice of prime mover is predominately electrical, but others have been used.

I.a.1.2 Condensers
The system is named according to the heat-transfer medium, either air or water. An air-cooled condenser, as the name implies, cools the hot compressed gas phase by forced air around the refrigerant heat exchanger. In a water-cooled system, the refrigerant condenser is cooled by water that can be sent directly to waste or recycled after itself being cooled in a direct or indirect water cooler. In a water cooling tower, hot condenser water is sprayed or otherwise distributed, over packings or pans that allow the water to trickle downward in thin films countercurrent to a rising airflow. Some of the heated water evaporates, extracting the heat of vaporization from the water and rejecting it to the ambient air. The water returns to the condenser cooler than when it left. In an evaporative cooler, water from a secondary source is sprayed over a closed loop coil in which the hot condenser water flows. Air is forced over the outside wetted coil surfaces to evaporate the secondary water, which, in turn, cools the primary condenser water inside the coil. One of the advantages

of this system is that the primary condenser water is not contaminated, but this system is slightly less efficient than the cooling tower configuration.

I.a.2 Expansion Devices

There are two commonly used expansion devices: thermostatically controlled expansion valves and capillary tubes.

I.a.3 Evaporators

The liquid refrigerant at low temperature is drawn into a low-pressure region by compressor suction where it evaporates, withdrawing the heat of vaporization from the air being cooled or from water used to cool the ventilation air. There are various kinds of evaporators in use, but the most widely used are refrigerant-to-air or refrigerant-to-water types.

I.a.3.1 Refrigerant-to-Air Evaporators

The liquid refrigerant is evaporated directly into a coil located inside an air handler that supplies cool air directly into conditioned spaces. In this case, a thermostatically controlled expansion valve is normally used.

I.a.3.2 Refrigerant-to-Water Evaporators

In this type of equipment the liquid refrigerant is evaporated into a coil that cools water surrounding the coil. The cooled water is then pumped through cooling coils located in the space to be cooled. This system is normally referred to as a "chiller." It is commonly used for large, widely distributed areas or buildings and when close control of environmental conditions is required.

I.a.3.3 System Descriptions and Strategies

After the cooling and heating equipment selection is made, the distribution system must be selected. There are three possible combinations: an all-air system, an all-water system, or a combination of the two.

Some commonly used combinations are:

1. *A reciprocating-type direct expansion system* consists of a reciprocating hermetic or semihermetic compressor connected to an air-cooled condenser followed by an expansion device and directly into a refrigerant-to-air coil in an air handler. A small version of this system is the common window air conditioner. Large systems of this nature are known as "package units," or "rooftop package units," because they are often

installed on roofs. A variation of this system is a so-called *split system,* where some of the components of the refrigeration cycle are inside and the heat rejection, or condenser, parts are outside.

2. *A chiller/cooling tower combination* is the most popular of the central refrigeration systems. A complete reciprocating or centrifugal compressor system is used to chill a secondary cooling medium (water) that is then pumped throughout the spaces to be cooled. The chilled water is incorporated into air handlers that consist of large housings containing fans, cooling coils, and filters. They cool the air and move it into the conditioned spaces. Heating coils are usually installed for winter heating and to reheat air for summer air conditioning that has been cooled below the comfort level for the purpose of condensing out excessive water vapor. These air handlers can be either low-pressure or high-pressure systems. A low-pressure system has less than 3 in. w.g. static pressure capability, whereas a high-pressure system exceeds this value.

I.a.3.3.1 All-Air Systems

In all-air systems, one or several air handlers are used to condition the entire building. A further classification can be made between:

1. *Constant Temperature Systems* in which air is cooled to a constant discharge temperature. Reheat coils to temper the air to any desired temperature are installed downstream to accommodate variations in requirements.
2. *Dual-Duct Systems* in which a single or multiple air handlers have a split air system, called a hot deck and a cold deck that simultaneously produces separate hot and cold airstreams. Both are distributed to each conditioned zone where a preset thermostatic controller positions dampers in a mixing box to mix the hot and cold airstreams to produce the desired air temperature in the room.
3. *Multizone Systems* are similar to dual-duct systems in that the air handler has a hot and a cold deck. However, the mixing zone dampers will be located at the air handler.
4. *Variable Volume Systems* supply air at a constant temperature into the conditioned space where the zone thermostat controls a volume box that reduces or increases the amount of air supplied to the conditioned space to satisfy the thermostat setting.

I.a.3.3.2 All-Water Systems

Chilled water for cooling, and in some cases hot water for heating, is piped throughout the conditioned space. In each conditioned space there are fan units to blow air over the coils to condition the space. The terminal fan-coil units are small, unitized cabinets with filters, a small fan, and coils for cooling and/or heating. All-water systems of this type are classified as:

1. *Two-Pipe Systems* in which a single set of water-containing pipes is distributed to each of the fan coils. The circulating water may be cool or hot, depending upon the season. Changeovers from hot to cold water and vice versa are conducted as the seasons require.
2. *Four-Pipe Systems* in which hot and cold water are supplied to each fan-coil unit through a separate set of pipes for instantaneous heating or cooling at each fan-coil unit as needed.

Whenever a terminal fan-coil unit is used, a condensate drain is normally needed to remove water that condenses on the outside of the cooling coil during humid weather.

I.a.3.3.3 Combined Systems

Mixed air and water systems are sometimes employed when special conditions or unusual requirements dictate their use, but they are less common.

ASHRAE Comfort Standards

Although the air conditioning engineer strives to satisfy the largest possible percentage of occupants, some may complain of being too hot while others complain of being too cold, no matter how well designed the system may be or how great an effort is made to satisfy everyone. This is because feelings of discomfort can occur even when the external environment is well within limits universally accepted as beneficial for enjoyment of life. Research on indoor comfort conditions has shown that various combinations of temperature, humidity, air movement, and heat radiation tend to produce equal feelings of comfort, and this has led to the development of a number of indices referred to as "effective temperature," "operative temperature," "equivalent temperature," "resultant temperature," and "equivalent-warmth index," each of which seems to account for the ways that combinations of these factors affect the subjective sensation of comfort.

Effective temperature combines indoor temperature, relative humidity, and air velocity into a single value that has been correlated with the results of empirically determined comfort evaluation polls. Values differ for men and women, for summer and winter, and for different areas of the continental United States and Canada; with somewhat higher temperatures being preferred in the north than in the south. As clothing habits and preferences of the affected populations have changed over the decades, the effective temperature charts have been modified accordingly (ASHRAE, 1981). Effective temperature is the most widely used comfort

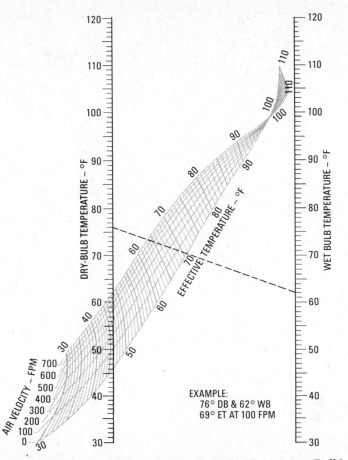

FIGURE II-1. Effective temperatures (at rest). From *Fan Engineering* (Buffalo Forge, 1983). From data of C. P. Yaglou and W. E. Miller, "Effective temperature with clothing" *Trans. ASHVE,* **31,** 1925, pp. 89–99.

index, perhaps because it is the oldest, having been published in 1925 (Yaglou, 1925). It has been criticized for neglecting radiant heat effects and for tending to magnify the warming effect of high-humidity conditions within the normal indoor temperature range. In spite of these defects, it remains a useful guide of comfort ventilation conditions for men and women at rest or engaged in light physical activity, the condition for which it is designed. Some of the other comfort indices are more appropriate for those engaged in hard physical labor or for work under high heat stress situations such as occur in foundries, smelters, and drop forge shops.

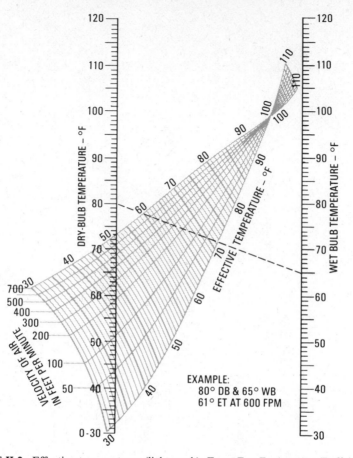

FIGURE II-2. Effective temperatures (light work). From *Fan Engineering* (Buffalo ᵀ 1983). From data of C. P. Yaglou, "Comfort Zones for Men at Rest and Strippᵉ Waist," *Trans ASHVE,* **33,** 1927, pp. 165–179.

Figures II-1 and II-2 are typical effective-temperature charts for people at rest and when engaged in light work that have been adapted from the original effective-temperature charts of Yaglou and Miller (Yaglou, 1925).

Figure II-3 is a version of the effective-temperature scale that has been adopted as Comfort Standard 55.74 by the American Society of Heating, Refrigerating, and Air Conditioning Engineers (ASHRAE, 1977). It applies to lightly clothed, sedentary individuals with low air movement, and avoids involvement with the effects of radiant energy by specifying that the mean radiant temperature (MRT) will be equal to air temperature.

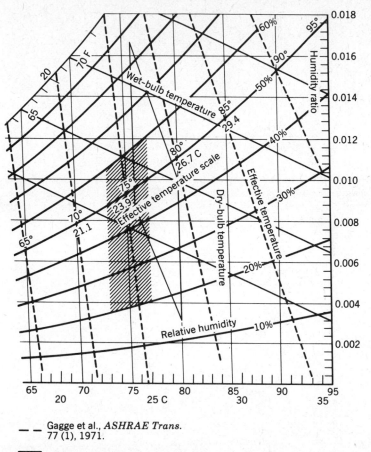

— — Gagge et al., *ASHRAE Trans.*
77 (1), 1971.

▨▨▨ *ASHRAE* Comfort Standard 55-74

FIGURE II-3. New effective temperature scale (ET). Envelope applies for lightly clothed, sedentary individuals in spaces with low air movement, where the MRT equals air temperature. With permission of the American Society of Heating, Refrigerating and Air-Conditioning Engineers, Inc., Atlanta, GA.

ASHRAE Comfort Standard 55–74 applies to the hatched area shown in the center of the chart in Figure II-3. Figure II-4 adapted from ASHRAE, 1977, shows the percentage of individuals who are predicted to experience dissatisfaction over a range of effective temperatures calculated from the chart shown in Figure II-3.

The comfort standard was revised in 1981 by ASHRAE as standard

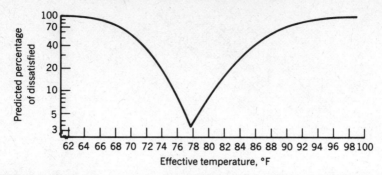

FIGURE II-4. Predicted percentage of individuals who would be thermally dissatisfied (PPD) at various levels of effective temperature (ET). Based on 1300 American and European subjects. With permission of the American Society of Heating, Refrigerating and Air-Conditioning Engineers, Inc., Atlanta, GA.

55–81 (ASHRAE, 1981). Considerably more research is on-going to further define, in a quantitative and qualitative manner, how human comfort is perceived.

APPENDIX III

Fans

III.a FAN TERMINOLOGY

The following definitions have been adapted from the 1983 edition of *Fan Engineering, An Engineer's Handbook on Fans and Their Applications* (Buffalo Forge, 1983).

FAN
: *Generally,* any device that produces a current of air by the movement of a broad surface.
: *Specifically,* a turbo machine for the movement of air having a rotating impeller at least partially encased in a stationary housing.

VENTILATOR
: A very-low-pressure-rise fan.

EXHAUSTER
: A fan used to remove air or gases from something or some place.

BLOWER
: A fan used to supply air or gases to something or some place.

IMPELLER
: The rotating element of a fan that transfers energy to the air (also called a wheel, rotor, squirrel cage, or propeller).

BLADES
: The principal working surfaces of the impeller (also called a vane, paddle, float, or bucket).

SHROUD
: A portion of the impeller used to support the blades

	(also called a cover, disk, rim, flange, inlet plate, back plate, or center plate).
HUB	The central part of the impeller that attaches to the shaft and supports the blades directly or through a shroud to the shaft.
HOUSING	The stationary element of a fan that guides the air before it enters the impeller and after it leaves the impeller (also called a casing, stator, scroll, scroll casing, ring, or volute). Centrifugal fan housings include side sheets and scroll sheets. Axial-fan housings include the outer cylinder, inner cylinder, belt fairing, guide vanes, and tail piece.
CUTOFF	The point of the housing closest to the impeller (also called the tongue).
INLET	The opening through which air enters the fan (also called the eye or the suction).
OUTLET	The opening through which air leaves the fan (also called the discharger).
DIFFUSER	A device attached to the outlet of a fan to transform kinetic energy to static energy (also called a discharge cone or evasé).
INLET BOX	A device attached to the inlet of a fan to make it possible to use a side entry into a centrifugal fan (also called a suction box).
VANES	Stationary blades used upstream or downstream of a fan to guide airflow. (When used upstream they are also called inlet guide vanes; when used downstream they are also called straightening vanes or discharge guide vanes.)
AXIAL-FLOW FAN	A fan contained in a cylindrical housing and characterized by flow through an impeller that is parallel to the shaft axis. *vane axial* fans have stator vanes. *tube axial* fans do not have stator vanes. *propeller* fans are pedestal or panel-mounted low-static-pressure axial-flow fans.
CENTRIFUGAL FAN	A fan contained in a scroll-shaped housing characterized by radially inward and outward flow through the impeller.

Typical centrifugal and axial-flow fans are illustrated in Figures III-1 and III-2.

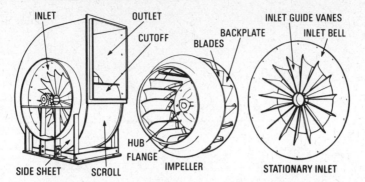

FIGURE III-1. Exploded view of a centrifugal fan. From *Fan Engineering* (Buffalo Forge, 1983).

III.b EXHAUST FAN SPECIFICATIONS

Centrifugal fans, complete with motor and drive, should be used for general exhaust service as well as for exhausting fume hoods. They should be of Class I construction, in accordance with the standards established by the Air Moving and Conditioning Association (AMCA, 1974). The fan impeller should be backwardly inclined and should have a true self-limiting horsepower characteristic. The fan should be single width, single inlet with ball bearing and overhang pulley for V-belt drive. In some cases the fan/motor drive combination should be an explosion-proof type. Fan housing should be constructed of steel and all parts should be bonderized and then coated with baked primer-finisher especially formulated to meet stringent corrosion-resistance standards. The coating should have a thickness of at least 1–2 mils without voids. The fan housing should be weatherproof for protection of motor and drive when located on the roof and a drain connection should be provided in the fan housing.

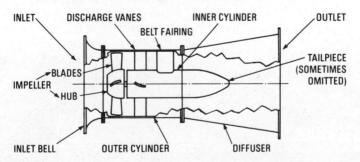

FIGURE III-2. Cutaway view of a vane-axial fan. From *Fan Engineering* (Buffalo Forge, 1983).

ASHRAE Filtration Guide

Performance of Throwaway Filters Used for Supply and Exhaust Air Cleaning

Filter Media Type	ASHRAE Weight Arrestance, %	ASHRAE Atmospheric Dust Spot Efficiency, %	MIL-STD 282 DOP Efficiency, %	ASHRAE Dust-Holding Capacity, Grams Per 1000 cfm/ Rated Capacity
Open cell foams and textile denier nonwovens	70–80	15–30	0	180–425
Thin, paperlike mats of glass fibers, cellulose	80–90	20–35	0	90–180
Mats of glass fibers, multi-ply cellulose, or wool felt	85–90	25–40	5–10	90–180
Mats of 5- to 10-μm fibers, 6 to 12 mm thickness	90–95	40–60	15–25	270–540
Mats of 3- to 5-μm fibers, 6 to 20 mm thickness	>95	60–80	35–40	180–450
Mats of 1- to 4-μm fibers	>95	80–90	50–55	180–360
Mats of 0.5- to 2-μm fibers (usually glass fibers)	NA[a]	90–98	75–90	90–270
Wet laid papers of glass fibers <1 μm diameter (HEPA filters)	NA[a]	NA[a]	99.99	500–1000

[a] NA, Indicates that test method cannot be applied to this level of filter.
Source. Adapted from ASHRAE, 1983.

Laboratory Hoods and Other Exhaust Air Contaminant-Capture Facilities and Equipment

GENERAL COMMENTS

Frequent references have been made throughout the text to laboratory hoods of various designs and to alternative exhaust systems that provide control of toxic, corrosive, combustible, and reactive gases, vapors, and aerosols originating from the materials being used and the processes employed. Assembling information on each of the options into a single section should help make those having design responsibility for these items better informed and thereby improve selection of this class of equipment. Because most of this equipment is built-in during construction or renovation, selection errors tend to become chronic sources of complaint after the facility is put into service and unusually costly to remedy. Therefore, thorough familiarity with laboratory hoods and all manner of laboratory contamination-control exhaust systems is essential for everyone involved in the laboratory design process. It cannot be overemphasized that laboratory workers are trained from their earliest introduction to experimental science to regard their laboratory hood as their principal, and all-purpose, safety device. Therefore, they tend, unquestioningly, to accept the efficacy of all such facilities. When the safety devices do not perform their intended function adequately, serious harm is likely to result. The laboratory designer's responsibility for providing facilities adequate to safeguard laboratory workers' health and safety is a heavy burden that cannot be delegated or evaded.

231

The sections that follow describe each of the exhaust ventilation contamination-control devices that are commonly found in laboratories. In spite of significant differences in design, they perform an equivalent function. Therefore, selection is generally guided by tradition, cost, energy consumption, and by similar factors, each of which may receive a more or less important rating by different laboratory users, designers, and owners. Regardless of the selection criteria used, it should be kept in mind that correct selection of blowers and duct construction materials and methods is essential for a full measure of safety and user satisfaction. The latter topics are treated in other sections: Appendix III for exhaust fans and Appendix VI for exhaust ducts and accessories.

V.a CONVENTIONAL BYPASS CHEMICAL FUME HOODS

V.a.1 Introduction

Bypass chemical fume hoods constitute the most widely used type of laboratory hood and are familiar to all laboratory workers. It is an improved version of an even older type that used to be referred to as a fume cupboard (a term still used in Britain) and, true to its name, the fume cupboard was an open-faced wood and glass, boxlike enclosure connected to an exhaust blower, usually located at the top rear. Many important modifications have been made to the basic design to improve airflow characteristics for the purpose of achieving greater fume-capturing efficiency and reducing energy requirements.

The most important modification to the old-fashioned fume cupboard was the addition of a vertical sliding sash covering the open front. The sliding sash made it possible to expose the entire open front for full access but also made it possible to lower the sash to provide a safety shield between experiment and experimenter. Lowering the sash during dangerous experiments also decreased the open-face area and this served to increase the inflow velocity through the remaining opening, thereby providing greater protection for the worker against an outflow of contaminants. However, when the sliding sash was lowered beyond a certain point the inflow velocity (for the same air volume required to exhaust-ventilate the hood when the sash was in the full open position) became so great that air turbulance effects at the edges of the opening caused backflow inside the hood, decreasing safety rather than enhancing it. In addition, excessively high velocities sweeping across the floor of the hood disrupted experiments by extinguishing burners, scattering powders, and cooling constant temperature flasks.

V.a.2 Important Features of Bypass Chemical Fume Hoods

To correct the several deficiencies of the old fume cupboard, the following features were developed and are found in modern, efficient bypass chemical fume hoods:

- Bottom and side airfoils around the open face to produce turbulence-free airflow into the hood.
- A mechanism to minimize excessive velocities (>300 fpm) when the total opening is 6 in. or less. This may be accomplished by an air bypass or by switching the fan to a lower speed. A typical bypass hood is illustrated in Figure V-1.

Other important characteristics of acceptable bypass chemical fume hoods are the following:

- The sash will be constructed of shatterproof material.
- The material of construction of the hood will be resistent to damage by the materials to be used in the hood.
- Non-asbestos-containing materials will be used.
- The hood will provide adequate containment. This can be evaluated by setting off a 30 s smoke candle or other heavy smoke generator inside the hood with the sashes in the fully open and then in the fully closed position. The smoke will be totally and quickly removed and there will

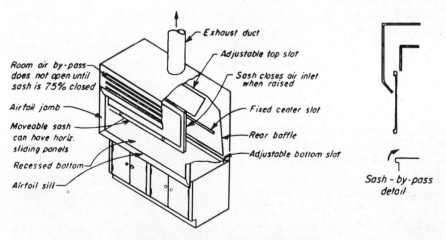

FIGURE V-1. Airfoil hood (ACGIH, 1984).

be no reverse flow out of the hood. Refer to Appendix Vh for other details on evaluating hood performance.

• There will be an inward airflow across the entire opening. Reverse flow can be detected by passing a smoke stick or equivalent smoke generator across the entire perimeter of the face opening and looking for flow out of the hood.

• No velocity measurement across the hood face with the sashes positioned to provide the maximum opening will be less than 80 fpm or greater than 120 fpm.

• The airflow through the hood will provide an average face velocity of 100 fpm at the maximum face opening.

• The airflow through the system will be monitored by an in-line measurement device (Pitot tube, orifice meter) or by static pressure measurements calibrated to the specific hood system. The dial of the measurement instrument will be visible to the hood user.

• During renovations, when existing fume hoods are to remain, an attempt to modify them should be made, if necessary. This would include providing a bottom airfoil, if missing and providing some means of controlling face velocity at all sash positions by means of a bypass or air volume control.

V.a.3 Model Hood Specification

To further illustrate the important characteristics of an acceptable bypass chemical fume hood, the *Laboratory Fume Hood Specifications and Performance Testing Requirements,* by the Environmental Medical Service at the Massachusetts Institute of Technology (Chamberlin, 1979) has been adapted as a model specification.

V.a.3.1 Standard Bypass Fume Hood

1. All fume hoods shall be of the airfoil design with foils at the bottom and along both vertical sides of the face opening. The bottom airfoil will be raised approximately 1 in. to allow air to pass under the foil and across the work surface and to serve as the terminus for vertical sliding sash(es). The vertical foils are to be flush with the hood interior surface to minimize turbulence as air enters the hood.

2. The superstructure construction for all hoods shall be counter mounted and not more than 65 in. high, 36 in. deep, outside dimensions, and of length indicated. Interior clear working height not less than 47 in. above work surface for full depth of hood from interior of lintel panel to face of baffle plenum. Double side wall construction is to enclose all

structural reinforcements, sash balance mechanisms, and mechanical connections for service outlets and controls (as indicated) and shall be of airfoil type not more than 4 in. wide. Access should be provided for the inspection–maintenance of the sliding sash operating mechanism and also for the installation of mechanical services.

3. The materials for the hood superstructure (unless otherwise specified) shall be $\frac{1}{4}$-in.-thick non-flammable, acid resistant material for the hood interiors and laboratory steel, factory painted (baked enamel) for the exterior. Color to be selected by the owner or architect from the manufacturer's standard colors (unless otherwise specified).

4. All hoods shall be of the "bypass" type. The bypass shall be located above the hood face opening and just forward of the sash when raised. All air exhausted must pass through the work chamber. The bypass must provide an effective line of sight barrier between the area outside the hood and the hood interior and must also provide an effective barrier capable of controlling transfer of flying debris during an explosion within the hood. It must assure essentially a constant volume of air at all sash positions. The by-pass shall control the increase in face velocity as the sash is lowered to at least twice the design velocity but not more than three times the design velocity.

5. All hoods shall have a vertical sliding glass sash of $\frac{7}{32}$-in.-thick combination safety sheet with metal frame to operate on stainless steel cables over the ballbearing pulleys with metal counter balancing weights. (*Note.* Spring sash balances are not acceptable.) Sash when in the closed position shall rest on top of the bottom airfoil; and when fully raised the height of the open face must be at least 32 in. from the top of the work surface. The sash must operate freely. Install a recessed finger grip(s) or drawer pulls (2 per sash) for raising and lowering sash.

6. The bottom airfoil shall be fabricated from stainless steel (316).

7. All hoods shall have a removable baffle with two slots; one upper and one lower, adjustable, with hand-operated plastic adjusting knobs for the adjustment of the airflow. A protective stainless steel screen with the equivalent free area of the total bottom slot opening shall be provided for the bottom slot. This protective screen shall not impede the use of the cup sink.

8. All hoods shall have an exhaust plenum chamber and be equipped with a hood outlet opening sized for approximately 1500 fpm at design flow. This opening shall be located to the rear of the back baffle and centered in top of exhaust plenum unless otherwise specified. The opening should be equipped with a stianless steel (316) duct stub extending at least 1 in. above top of hood.

9. Hood work surface shall be type 316 stainless steel and shall be of the recessed (dished) type with a $\frac{3}{8}$-in. raised lip along all four edges (to retain spilled liquids) and a uniform edge thickness of $1\frac{1}{4}$-in. The work surface shall contain an integrally welded type 316 stainless steel 3 × 6-in. cup sink located near one of the rear corners unless otherwise specified, so that it is not obstructed by the protective screen at the bottom slot and will also receive the unobstructed discharge stream from the cold water outlet.

If the storage cabinet is to be vented, the raised surface of the work top, above the recessed area, shall contain one or two holes to receive the $1\frac{1}{2}$-in. I.D. lead vent pipe(s) from the acid storage cabinet(s) supporting the hood. Vent pipe holes shall be located between the rear baffle and back panel of the hood, in the raised portion of the work top. The raised surface shall be provided all around the recessed pan area, and it shall be 2 to 4 in. wide across the front edge.

10. At the top and front of each hood install a pocket enclosure to receive the vertical sliding sash when in the up position. Each pocket enclosure may contain two removable access panels—one each on the front and rear faces—for access to the electrical junction box and lighting fixture (relamping and cleaning). Provide a removable end panel on each side of the hood, extending from the sash enclosure to the rear of the hood, for access to service fittings and service lines. The panels must be removable from the exterior of the hood, and re-installable from the exterior, without interference with or removal of any drop ceiling, furring from the top of the hood to the ceiling, or any adjoining end curbs of work tops.

11. All fume hoods shall be equipped with $\frac{1}{4}$-in.-thick non-flammable, acid resistant material, flush, removable access panels(s) on the side(s) of hood interior of sufficient size required for the installation of the various mechanical services.

12. The hoods shall be exhausted so as to maintain an average velocity of 100 fpm through the full open face (sash fully raised). The velocity must also be uniform and not vary more than ± 10 fpm from the average.

13. When two speed exhaust fan systems are used, the microswitch that will control the exhaust air fan motor shall be actuated by hood sash only. The microswitch shall be so located that when the exhaust is turned to low speed the design volume of air is still exhausted from the room while maintaining a minimum face velocity of 100 fpm through the reduced face area.

14. *Note.* To the electrical and plumbing subcontractors and hood manufacturers: Unless otherwise noted on drawing or job specification,

piping from the hood service outlets, except for the drain and vent lines, shall be installed between the double wall side walls and extend up and be connected to the respective services *above* the hood.

15. All hoods shall have a fluorescent lighting fixture operated by an exterior switch mounted on exterior face of one of the vertical foils. The fixture shall have two 40-W, fluorescent, rapid start lamps. The fixture shall be hinged on one side and be accessible for relamping, cleaning, or other maintenance work from the exterior of the work chamber. Mount the fixture at top of hood, setting it on a fixed and gasketed $\frac{1}{4}$-in.-thick safety glass shield.

16. All hoods shall have one or more 15-A duplex grounded receptacle(s) mounted on the exterior face(s) of the vertical foil(s). The opening and/or housing for the receptacle should have a built-in through-flow ground fault interruptor (GFI).

Note. The wiring within the hood shall be so installed that when a through-flow GFI is installed it also gives GFI protection to the lighting switch and lighting fixture. When a hood has more than one duplex receptacle, only one through-flow GFI is required, which is to give protection to any other receptacle in the hood.

17. The lighting fixture with its operating switch and the electrical receptacle(s) for each hood shall be on the same circuit. The wiring for these electrical items shall connect into a single 4 × 4-in. junction box so located above the hood structure that it is always easily accessible no matter in what position the hood is placed.

V.a.3.2 Hood Support Cabinets

1. Base cabinet for hood support shall be constructed and finished in a manner similar to hood.

2. Provide acid storage cabinets as hood support cabinet. Interior of cabinet shall be completely lined with $\frac{1}{4}$-in.-thick non-flammable, acid resistant material including the interior sides of cabinet doors.

3. Each base cabinet shall be vented if used for storage of toxic materials. Each individual door shall have an air intake louver located at center and bottom of door. Each base unit shall be exhausted by means of a $1\frac{1}{2}$-in. I.D. lead pipe vent that extends from the center and top of the rear wall (unless otherwise noted) of the storage compartment itself and up through the 2-in. I.D. hole in the raised portion of the work surface (refer to Section V.a.3.1, item 9) and into the hood (behind the rear baffle plate) and terminating a minimum of 2 in. above the work surface.

4. Provide a 12-in. deep, full-width, noncombustible, and adjustable

shelf in each base unit. The shelf must be of sufficient strength or depth so that excessive deflection does not occur when it is fully loaded.

V.a.4 Horizontal Sliding Sash Option

Although most chemical fume hoods are provided with vertical sliding sash, it is possible to utilize horizontal sliding sash, instead. The principal advantage of horizontal sliding sash is that only half of the hood face can be left open at any time while giving full access to all parts of the hood interior. With vertical sliding sash, it is necessary to open the sash to its full height to gain access to the upper parts of the hood for equipment setups and monitoring tall apparatus. A hood with horizontal sliding sash is illustrated in Figure V-2. With horizontal sliding sash, the full height of the opening is always available whenever the sash is open, facilitating ready hood access. Additional details of the horizontal sliding sash modifications for the chemical fume hood are contained in Chapter 21, concerned with energy conservation, because use of horizontal sliding sash is an important method for reducing the need for laboratory exhaust air volume.

V.b AUXILIARY AIR CHEMICAL FUME HOODS

V.b.1 Introduction

Auxiliary air chemical fume hoods differ from bypass chemical fume hoods in only one important respect: a major portion of the air exhausted

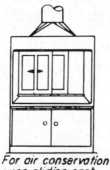

For air conservation
use sliding sash

Q = 100 - 150 cfm/sq ft open door area
Entry loss = 0.5 VP
Duct velocity = 1000 - 2000 fpm
          to suit conditions

FIGURE V-2. Hood with horizontal sliding sash (ACGIH, 1984).

from the hood is provided from a supply air diffuser canopy attached to the hood just above the hood face. Figure V-3 is an illustration of an auxiliary air chemical fume hood. The purpose of this modification of a conventional bypass chemical fume hood (Appendix V.a) is to reduce the demand for fully conditioned makeup air for hood service (see Chapter 21 on energy conservation). This is accomplished during cold weather by heating the auxiliary air to a lower temperature than the room HVAC supply air, and during hot weather, by not cooling or dehumidifying the auxiliary supply air. The rationale is that the hood user ordinarily spends little time working in front of the hood and, in any event, will not be seriously discomforted by unconditioned air flowing from above into the open hood face. In hot climates, especially, the savings that can be realized by not cooling and dehumidifying all the hood exhaust air can be substantial. However, this type of chemical fume hood requires an auxiliary air supply system for each hood that is separate from the room HVAC supply air system, thereby increasing the number and complexity of the HVAC services to the laboratory.

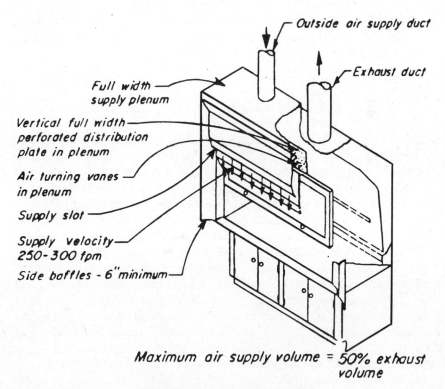

FIGURE V-3. Air-supplied hood.

V.b.2 Model Hood Specification

The requirements for auxiliary air chemical fume hoods that differ from, or are supplementary to, those outlined in Appendix Va are the following by way of illustration. They are adapted from a hood specification schedule used by the Environmental Medical Services of the Massachusetts Institute of Technology for hood procurement (Chamberlin, 1979):

1. All of the performance and construction requirements for bypass chemical fume hoods must be met. (Refer to Appendix V.h for performance tests.)

2. The supply plenum shall be so located that all of the auxiliary air shall be supplied exterior to the hood face and no leakage or passage of auxiliary air behind the sash will be allowed until the sash is lowered to the point of bypass opening. This system shall be capable of supplying up to 70% of the volume of the hood exhaust.

3. The supply air shall be tempered to 60°F during the "winter" season.

4. The capture efficiency of the supply air by the hood exhaust system shall be at least 95% within the supply air temperature range of 60 to 90°F.

5. The fume hood leakage within the above operating range shall not exceed 0.0003% as determined by performance tests.

V.c PERCHLORIC ACID FUME HOODS

Perchloric acid hoods are special because perchloric acid digestions volatilize the entire volume of perchloric acid that is used; that is, the digestions are conducted to dryness. Perchloric acid is a powerful oxidizer. It is a high-boiling chemical that undergoes spontaneous and explosive decomposition. The same is true of many perchlorate salts. In conventional hood exhaust ducts, volatilized perchloric acid cools, condenses, and deposits in horizontal runs, creating, in time, a severe explosion hazard. To avoid deposition and accumulation of perchloric acid in hood exhaust ducts, perchloric acid hood systems have been developed. They feature dedicated duct systems that have few, if any, horizontal runs, no dead areas that can capture and accumulate condensed perchloric acid, and that are constructed of materials such as stainless steel with welded seams or neoprene gaskets at joints that do not degrade under perchloric acid attack. In addition, they should be located as close to the outside of the building as possible and have stainless steel fans.

The additional special characteristics of perchloric acid hood systems are that the interior of the hood is also constructed of materials that are not degraded by perchloric acid and that all of the exhaust ducts are continually water washed into a sewer or sump to remove condensed perchloric acid and prevent accumulation of the acid or its salts. Perchloric acid hoods must meet the same performance requirements as those outlined in Appendixes V.a and V.b for bypass and auxiliary air chemical fume hoods.

V.d HOODS FOR WORK WITH RADIOACTIVE MATERIALS

Some radioactive materials, such as the long-lived α emitters, for example, plutonium, have exceptionally low permissible exposure limits and, consequently, hoods designated as radioactive hoods are usually operated at somewhat higher face velocities than chemical fume hoods. Many users of radioisotopes recommend 150 fpm average face velocity for hoods but most industrial hygiene engineers believe that this face velocity is too high because it promotes excessive air turbulence at the edges of the working opening. Instead, they recommend a maximum average face velocity of 125 fpm. Whether 125 or 150 fpm face velocity is specified, the hoods that are purchased for service with radioactive materials should include all the features outlined in Appendixes V.a and V.b. In addition, radioactive hoods should be equipped with 316, 18-gauge stainless steel ducts. All the interior hood surfaces should be constructed of a smooth, cleanable, non-porous material such as stainless steel.

Provisions should be made for HEPA filters and/or activated charcoal adsorbers to be installed at the hood air outlet when required by NRC regulations (NRC, 1981) and the fan should be selected to handle the increased static pressure produced by the air filtration system. Finally, provisions should be made to install a continuous effluent air radioactivity monitoring system when this is required by NRC regulations.

V.e GLOVE BOXES

When the toxicity, radioactivity level, or oxygen reactivity of the substances under study is too great to permit safe operation in a chemical fume hood, resort must be made to a totally enclosed, controlled-atmosphere glove box. The special feature of a glove box, as the name suggests, is the total isolation of the interior of the box from the surrounding environment and the consequent need to manipulate items inside the box by means of full-length gloves sealed into a sidewall of the box. To prevent loss of materials from the inside of the glove box to the laboratory, the box is maintained under substantial negative pressure (0.25 to 0.50 in. w.g.) relative to the laboratory by means of an exhaust blower connected to the box interior. The atmosphere inside the box may be maintained sterile and dust-free by use of a constant air in-leakage through a HEPA filter. As shown in Figure V-4, a diagram of a typical glove box, the interior may be further isolated from the laboratory environment by the use of air locks for passing items into and out of the box.

Stainless steel and safety glass are the preferred materials of construction for the interior of the box to facilitate cleaning and decontamination. The interior should be finished smooth and free of sharp edges that could damage the gloves. All controls should be located outside the box for safety and ease of manipulation, as shown in Figure V-4. When highly toxic or infectious materials are being used in the glove box, the exhaust air should be cleaned in two or more stages. First, a prefilter to remove the major load of coarse particulate matter followed by a HEPA filter with a minimum efficiency of 99.97% for 0.3-μm test aerosol particles. When highly toxic volatile chemicals are used inside the box, an activated charcoal adsorber stage may be added to the air cleaning train. If no toxic aerosols are present, the HEPA filter may be omitted but it is advisable to retain the prefilter to protect the activated charcoal from dirt. The specific size of the air cleaning components will depend on the glove box size and the chemicals used. The blower must be selected to overcome the flow resistance of these added elements.

V.f BIOLOGICAL SAFETY CABINETS

Three classes of biological safety cabinets are recognized (NSF, 1983):

CLASS I. A ventilated cabinet for personnel and environmental protection, with an unrecirculated inward airflow away from the operator. The cabinet exhaust air is treated to protect the environment before it is discharged to the outside atmosphere. This cabinet resembles a chemical fume hood with a filtered exhaust and is suitable for work with low- and moderate-risk biological agents where no product protection is required. It is not widely used.

CLASS II. A ventilated cabinet for personnel, product, and environmental protection having an open front with inward airflow for personnel protection, downward HEPA-filtered, laminar airflow for product protec-

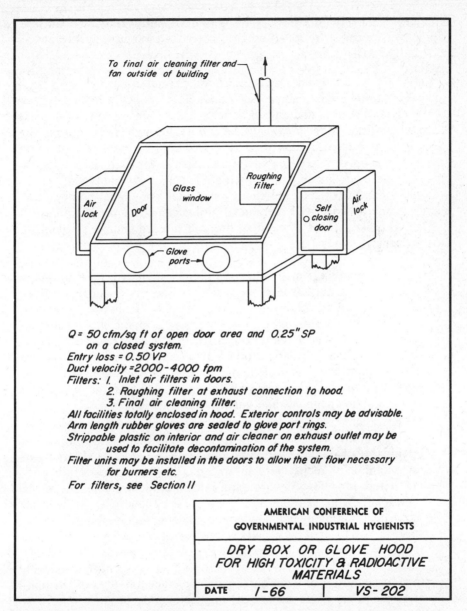

To final air cleaning filter and
fan outside of building

Roughing
filter

Glass
window

Air
lock

Door

Self
closing
door

Air
lock

Glove
ports

Q = 50 cfm/sq ft of open door area and 0.25" SP
      on a closed system.
Entry loss = 0.50 VP
Duct velocity = 2000 - 4000 fpm
Filters: 1. Inlet air filters in doors.
          2. Roughing filter at exhaust connection to hood.
          3. Final air cleaning filter.
All facilities totally enclosed in hood. Exterior controls may be advisable.
Arm length rubber gloves are sealed to glove port rings.
Strippable plastic on interior and air cleaner on exhaust outlet may be
      used to facilitate decontamination of the system.
Filter units may be installed in the doors to allow the air flow necessary
      for burners etc.
For filters, see Section II

AMERICAN CONFERENCE OF GOVERNMENTAL INDUSTRIAL HYGIENISTS	
DRY BOX OR GLOVE HOOD FOR HIGH TOXICITY & RADIOACTIVE MATERIALS	
DATE 1-66	VS-202

FIGURE V-4. Glove box. From American Conference of Governmental Industrial Hygienists (ACGIH, 1984).

tion, and HEPA-filtered exhaust air for environmental protection. Class II cabinets are suitable for low- and moderate-risk biological agents.

CLASS III. A totally enclosed, ventilated cabinet of gas-tight construction. Operations in the cabinet are conducted through attached rubber gloves. The cabinet is maintained under negative air pressure of at least 0.5 in. w.g. Supply air is drawn into the cabinet through HEPA filters. The exhaust air is treated by double HEPA filtration, or by HEPA filtration and incineration. Class III cabinets are suitable for high-risk biological agents and are accompanied by much auxiliary safety equipment. There are only a handful of such facilities worldwide. They must be designed, installed, and certified by experienced biological safety professionals.

Class II biological safety cabinets are widely used and are available in four recognized types:

Type A. Cabinets (1) maintain a minimum calculated average inflow velocity of 75 fpm through the work area access opening; (2) have HEPA-filtered downflow air from a common plenum (i.e., a plenum from which a portion of the air is exhausted from the cabinet and the remainder supplied to the work area); (3) may exhaust HEPA-filtered air back into the laboratory; and (4) may have positive-pressure contaminated ducts and plenums. Type A cabinets are suitable for work with low- to moderate-risk biological agents in the absence of volatile toxic chemicals and volatile radionuclides.

Type B1. Cabinets (1) maintain a minimum (calculated or measured) average inflow velocity of 100 fpm through the work area access opening; (2) have HEPA-filtered downflow air composed largely of uncontaminated recirculated inflow air; (3) exhaust most of the contaminated downflow air through a dedicated duct exhausted to the atmosphere after passing through a HEPA filter; and (4) have all biologically contaminated ducts and plenums under negative pressure, or surrounded by negative-pressure ducts and plenums. Type B1 cabinets are suitable for work with low- to moderate-risk biological agents. They may also be used with biological agents treated with minute quantities of toxic chemicals and trace amounts of radionuclides required as an adjunct to microbiological studies if work is done in the direct-exhausted portion of the cabinet or if the chemicals or radionuclides will not interfere with the work when recirculated in the downflow air.

Type B2. Cabinets (sometimes referred to as "total exhaust") (1) maintain a minimum (calculated or measured) average inflow velocity of 100 fpm through the work area access opening; (2) have HEPA-filtered downflow air drawn from the laboratory or the outside air (i.e., downflow air is not recirculated from the cabinet exhaust air); (3) exhaust all inflow and downflow air to the atmosphere after filtration through a HEPA filter without recirculation in the cabinet or return to the laboratory room air; and (4) have all contaminated ducts and plenums under negative pressure or surrounded by directly exhausted (nonrecirculated through the work area) negative-pressure ducts and plenums. Type B2 cabinets are suitable for work with low- to moderate-risk biological agents. They may also be used with biological agents treated with toxic chemicals and radionuclides required as an adjunct to microbiological studies.

Type B3. Cabinets (sometimes referred to as "convertible cabinets") (1) maintain a minimum (calculated or measured) average inflow velocity of 100 fpm through the work access opening; (2) have HEPA-filtered downflow air that is a portion of the mixed downflow and inflow air from a common exhaust plenum; (3) discharge all exhaust air to the outdoor atmosphere after HEPA filtration; and (4) have all biologically contaminated ducts and plenums under negative pressure or surrounded by negative-pressure ducts and plenums. Type B3 cabinets are suitable for work with low- to moderate-risk biological agents treated with minute quantities of toxic chemicals and trace quantities of radionuclides that will not interfere with the work if recirculated in the downflow air.

The above descriptions were taken from National Sanitation Foundation Standard No. 49 (NSF, 1983). Standard No. 49 includes basic requirements for construction and certification testing of all Class II biological safety cabinets. The appearance of the words "laminar flow" in the title of the standard makes it necessary to caution against confusing Class II biological safety cabinets with laminar flow workbenches because, although the latter devices provide work protection, they fail to provide personnel or environmental protection, and therefore should never be used with toxic, infectious, or otherwise hazardous materials. Figures V-5 and V-6 identify the important parts and show the airflow patterns for Type A and Type B1 cabinets, the two most widely used types.

The heart of every cabinet is the HEPA filter, a pleated paper filter contained inside a rectangular wooden frame. The filter is clamped tightly to a specially prepared flange built into the cabinet frame, and a compressible gasket on the face of the wooden filter frame makes the required airtight seal. Threaded clamps are usually used but some manufacturers use spring-loaded clamps.

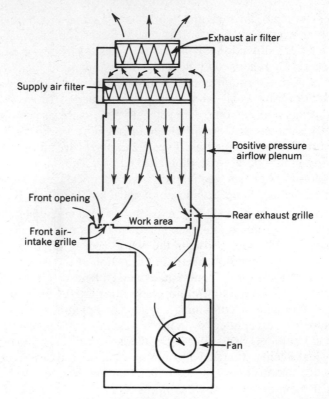

FIGURE V-5. Airflow patterns in NIH-03-112C Class II Type A biological safety cabinet.

Most cabinets use direct-drive blowers with forward curved impellers driven by permanent split-capacitor motors and controlled by a Triac speed controller. The Triac is a solid-state device that can be adjusted to vary the ac voltage at the output from essentially zero to full line voltage. The permanent split-capacitor motor is a brushless electric motor of fractional horsepower size. It requires an external capacitor that remains in the circuit during starting and while running. It is the largest ac motor that will run at variable speed, reducing rpm and horsepower as input voltage is reduced.

The forward curved blower impeller has a performance characteristic that prevents it from overloading the motor as the filters increase in resistance because the torque required to turn this blower increases rapidly with increasing airflow but only very little with increasing pressure. Therefore, as filter resistance rises, airflow tends to drop and the load on the motor to decrease, but this causes the motor to speed up and airflow

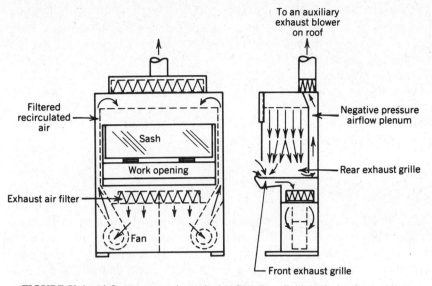

FIGURE V-6. Airflow patterns in typical NCI-I Type B biological safety cabinet.

remains close to its original value. Starting with new filters in a cabinet, airflow will not decline more than 10% even after filter pressure drop has increased 50%. But as filters increase above 50% of new resistance, the amount of pressure increase needed to produce a 10% drop in airflow decreases and more frequent manual adjustments are needed.

Use of UV lamps in biological safety cabinets was discontinued in the 1983 revision of NSF Standard No. 49 because of their limited effectiveness for decontaminating cabinets. Correct procedure, when using a biological safety cabinet, calls for washing down the work area with a suitable disinfectant upon completion of work. Without a good washdown, remaining soil is likely to shield organisms from the UV light, whereas with a good washdown, further disinfection of the cabinet surfaces is generally not needed; especially if the blower is left running continuously.

All cabinets contain electrical utility outlets. Some are useful only for powering small appliances whereas others have full 15-A capacity. All units should have ground fault interrupters in the electrical utility outlet line. They measure leakage to ground and automatically cut off power when leakage exceeds 5 mA. The proper functioning of the ground fault interrupter must be verified for all units whenever cabinet certification tests are performed.

Prior to sale, cabinets are submitted for certification to the National Sanitation Foundation by the manufacturers. Those that have passed the personnel, environmental, and product safety tests can be identified by a distinctive NSF medallion placed on the exterior of the cabinet. Field recertification by a competent technician is needed when a cabinet is first installed, at annual intervals thereafter, and whenever a cabinet is moved to a new location or is serviced internally. It is standard safety practice to sterilize the entire internal structure of a working cabinet with formaldehyde gas each time it is necessary to service interior parts. NSF Standard No. 49 contains detailed instructions for performing the field certification tests and for formaldehyde sterilization.

V.g CAPTURE HOODS

The most efficient and cost effective form of contaminant control is local exhaust ventilation (LEV). This involves capture of the chemical contaminant at its source of generation. The laboratory chemical fume hood (Appendix V.a) is a specialized form of capture hood that totally encloses the emission source. Often, total enclosure of the source is not possible, or is not necessary. A capture hood controls the release of toxic materials into the laboratory by capturing or entraining them at or close to the source of generation, usually a work station or laboratory operation. Considerably less air volume is required than for the standard chemical fume hood.

To work effectively, the air inlet of a capture hood must be placed near the point of chemical or biological release. The distance away will depend on the size and shape of the hood and the velocity of air at the intake slot or face, but should usually be not more than 12 in. from the generation source. Design face or slot velocities are typically in the range of 500–1500 fpm. Many design guidelines exist for this class of exhaust hoods in the Industrial Ventilation Manual (ACGIH, 1984). Two types of capture hoods that find frequent application in the laboratory are "canopy" hoods and "slot" hoods.

Canopy hoods are used primarily for capture of gases, vapors, and aerosols released from permanent laboratory equipment. Examples include ovens, gas chromatographs, autoclaves, and atomic absorption spectrophotometers. They are usually found in analytical and biological laboratories and some pilot plant operations. Many equipment manufacturers recommend specific capture hood configurations that are suitable for their units. An example can be seen in Figure V-7.

Slot hoods are used for control of laboratory bench operations that cannot be performed inside a containment hood (chemical fume hood) or under a canopy hood. Laboratories that may find use for slot ventilation

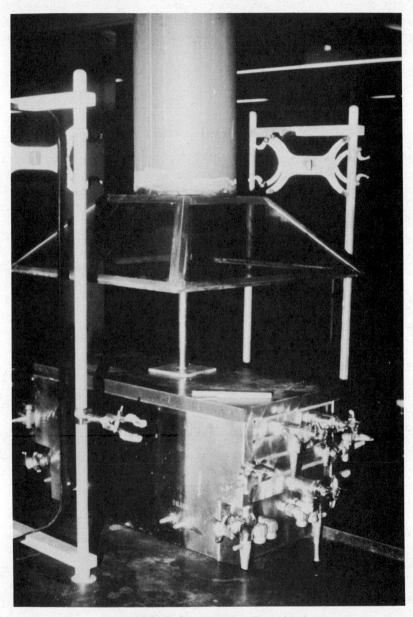

FIGURE V-7. Canopy-type exhaust hood.

include clinical (histology, pathology), anatomy, teaching, and general chemistry units. Typical operations include slide preparation, microscopy, biological specimen preparations, mixing, and weighing operations. An example is shown in Figure V-8.

V.h PERFORMANCE TESTS

V.h.1 Introduction

From the initial selection of the hood to its continuing use in the laboratory the owner will need some method of evaluating its performance. The method involves selecting the conditions under which the hood will be tested, choosing an appropriate challenge test, and developing acceptance criteria that will adequately determine if the hood meets the user's protection needs. There are a variety of performance tests currently used. We will discuss several, indicating their appropriateness for specific applications and their advantages and disadvantages. Each user must determine which test will accurately determine if a specific hood will provide the needed protection.

FIGURE V-8. Bench-type slot hood. Note: Location of electrical outlets are not recommended.

Factors to consider when choosing a performance test include (1) reason for testing, (2) type and quantity of chemicals or biological agents to be used in hood, (3) types of operations and equipment to be used in hood, (4) number and type of users, (5) diversification of hood use, both in the short term (months) and long term (years), (6) location of hood within the facility, (7) type of hood (conventional or auxiliary air), and (8) ease of performance of test.

The conditions under which one would test a hood's performance characteristics normally fall into two categories: (1) selection of the type of hood to be purchased, and (2) evaluation of hoods in use within the facility. The first category involves testing in a controlled, laboratory-type setting, whereas the second involves an evaluation in the laboratory under practical use conditions. Both conditions will be discussed here.

Performance tests involve measurement of the hood flow characteristics (face velocity and air quantity) and the efficiency of the hood in containing an artificial challenge gas or aerosol generated within the hood.

V.h.2 Tests for Selection of a Hood

Prior to purchasing one or more hoods, the user must determine under what conditions the hood may be used, what type of hood to select (bypass vs. auxiliary air), and what protection factor is needed, for example, what is the maximum allowable loss of containment (leakage) that is acceptable. (Refer to Section 2.3.4.3 for assistance in defining hood use and types.)

Considerable research on performance tests has been conducted since 1970 and several test protocols have been proposed for adoption as a standard. They involve a variety of challenge chemicals, methods of challenge, and criteria for acceptance. Some of the proposed challenge chemicals are uranine dye (Chamberlin, 1979), Freon (ASHRAE, 1984), and sulfur hexafluoride (Chamberlin, 1982). These tests have been employed under various conditions by several organizations. The reader is encouraged to review the above references in detail before making a selection. In addition, the current literature should be searched since this area is receiving considerable research attention.

V.h.3 Field Performance Tests for Fume Hoods after Installation

After hoods have been installed but before their use, they should be tested to verify adequate performance. Generally, this involves measurement of total volume flow and face velocity across the hood opening and comparison to design guidelines (see Section 2.3). Face velocity measurements should be made at 9 to 12 points equally distributed across the opening of the hood (Chamberlin, 1982; BSI, 1979; SAMA, 1975). In addition observation of airflow patterns should be made by generating a source of smoke across the face opening. It has also been common practice to conduct these tests at regular intervals throughout the year on opeational hoods.

For many years health and safety professionals have recognized that this procedure may not be the most accurate field assessment of hood performance. More recently, other field techniques have been attempted using some variation of the ASHRAE test referred to in Appendix V.h.2 (Fuller, 1979; Mikell, 1981). While these can provide much information after installation and before use, they are difficult to use on a regular basis since they are time consuming and can intrude on operations performed in the hood. A new method has been developed to provide an easy, quick and unobtrusive assessment of fume hood performance (Ivany, 1986). This test involves the release of a tracer gas through a diffuser inside the hood under normal operating conditions. Measurements of the tracer gas are made outside the hood to determine actual hood leakage. This leakage can then be related to chemical exposure standards. This test can be used as an after-installation/before-use acceptance criterion and as a regular interval performance test.

Whichever performance test is selected it is important to include the requirements in the design documents with criteria for acceptance.

APPENDIX VI

Exhaust Air Ducts and Accessories

Fume hood and capture hood ducts differ from HVAC ducts in that the materials that pass through them are often highly corrosive and toxic. Consideration should be given to the fact that such ducts will have to be serviced or replaced during the life of the average laboratory building. Therefore, safety to personnel making repairs or replacement should not be overlooked.

There is no concensus on the best material to use for exhaust duct construction. The following is a survey of a number of code compliance requirements.

BOCA (1985) specifies steel ducts for the removal of dust and vapors.

NFPA-90A specifies that ducts shall be constructed of steel, aluminum, or other inert incombustible materials. In addition, duct materials meeting the requirements of UL181, Class 0 and 1 are acceptable (UL, 1984). Class 0 materials have a surface-burning characteristic of zero. Class 1 materials have a flame-spread rating not over 25 and a smoke-developed rating not over 50.

ASHRAE (1982) recommends that exhaust ducts serving hoods which exhaust radioactive materials, volatile solvents, and strong oxidizing agents (perchloric acid) be fabricated of stainless steel for a minimum distance of 10 ft from the hood outlets.

SMACNA (1974) recommends that NFPA-91 and local building codes be consulted for fire resistance, fire-spread ratings, and smoke-generating characteristics. NFPA-91 applies only to the transmission of nonflammable fumes and vapors.

Many chemicals encountered in laboratories are flammable. For this reason, many fire departments oppose the use of rigid PVC ductwork due to the possibility of forming toxic fume degradation products in a fire. Transite has come into disfavor as a duct material because of its asbestos

TABLE VI-I
Commonly Used Duct Materials

Material	Limitations of Use
Glazed ceramic pipe	Rarely used today because of installation and sealing difficulties.
Epoxy-coated stainless steel	Extensive experience not yet available but appears promising as a versatile material.
Stainless steel	May be attacked by some chemicals, especially hydrochloric acid. Care should be used in selecting type. Stainless No. 316 is one of the more resistant alloys.
Monel metal	May be attacked by some chemicals, such as halide salts and acids.
Synthetic or cementitious "stones"	Absorb moisture, attacked by strong alkaline chemicals.
Reinforced plastic, principally glass fiber-reinforced polyester (FRP)	Various resins being used have different chemical and fire resistances. Care should be taken to select chemically resistant resins for final interior layer.
Aluminum	Limited resistance to many chemicals. Care should be used. Install only in systems that do not experience corrosive chemical exposure.
Galvanized steel	Limited resistance to corrosion by a wide variety of materials used in research and teaching laboratories. Not recommended.
Black steel	Useful only with dry and noncorrosive dusts.

content, but it is not clear that transite sheds respirable asbestos fibers during normal duct usage. Table VI-1 is a tabulation of commonly used duct materials.

In general, high-alloy stainless steels (e.g., 316) and glass fiber-reinforced polyester (FRP) have proven the most satisfactory duct materials for corrosion, impact, and vibration resistance, as well as ease of fabrication and installation. Rigid polyvinylchloride (PVC) has excellent corrosion resistance but is brittle and, therefore, has inferior impact and vibration-cracking resistance. Because plastic ducts have thicker walls than stainless steel of the same inside diameter, they occupy somewhat more space in duct and utility chases. Rigid PVC is nonflammable by reason of its high chlorine content, but may produce hydrogen chloride gas as a fire degradation product. Polyester resins are free of chlorine but can be formulated with phosphate and other fire-retardant additives that make them self-extinguishing, and this type of polyester should be insisted upon for all FRP ducts.

Considerable care must be exercised to evaluate the types of materials that are likely to be used in laboratories prior to making a selection of duct materials. Similar considerations will be a guide to the features that will be provided in the design of the building for ease of servicing and replacing ducts. It should be kept in mind, as well, that laboratory usage may change over a period of time and that considerable conservatism should, therefore, be exercised when selecting chemical fume hood duct materials to avoid inadvertant future failures.

High-velocity air movement in ducts is desirable to ensure that solids do not deposit in the joints, cracks, or corners of the duct system. A minimum suggested design velocity is 2000 fpm. Higher conveying velocities (3500–4500 fpm) are usual. A minimum of turns, bends, and other obstructions to airflow is also desirable. Where perchloric acid is to be used, the duct configuration should be free of bends and horizontal runs and permit thorough washdown of all interior duct surfaces.

A typical duct specification is likely to include some or all of the language that follows. "All fume hood and local exhaust ducts shall be constructed of round piping with the interior of all ducts smooth and free of obstructions. All joints shall be welded or epoxy sealed airtight. The use of flexible piping for spot exhaust points shall be kept to minimum lengths and shall be equipped with tightly fitting, easily removable end caps for use when the exhaust point is in use. Flexible tubing shall be noncollapsible and constructed entirely of metal or of a wire coil covered with multiple plies of flame-proofed, impervious fabric."

VI.a DUCT ACCESSORIES

VI.a.1 Dampers and Splitters

When dampers are used, they should be equipped with indicating and locking quadrants and the damper blades should be riveted to the supporting rods. Dampers should be of the same material as the ducts in which they are installed but two gauges heavier. Cast or malleable brackets, riveted to the sides of the duct, should be sufficiently long to extend full width of the branch ducts to which they are attached.

Opposed blade dampers should have each blade sealed with foam rubber or felt in order to form an effective seal between blades with the damper in the fully closed position.

VI.a.2 Fire Dampers and Fire Stops

Ten-gauge galvanized steel sleeve-type horizontal and vertical fire dampers should be installed in all ducts that penetrate fire walls or floors. The assembly should consist of 18-gauge galvanized, formed-steel blades with interlocking joints to form a continuous steel curtain when closed. The assembly should have a maximum depth of 4 in. and be suitable for horizontal or vertical airflow as required. Fire and smoke control dampers should meet or exceed NFPA 91 (NFPA 91, 1985). Fire dampers installed in ducts with either dimension under 12 in. should be constructed and designed so that the blade stock in the open damper position will be completely outside the airstream.

Unless required by code we do not recommend the use of fire dampers in hood exhaust ducts.

VI.a.3 Sheet Metal Access Doors

Hinged-type sheet metal access doors are needed in the ductwork at each automatic damper and control device and at each fire damper to give access to the fusible link. Access doors should be of the same material as the ductwork on which they are installed and constructed to be sealed airtight.

VI.a.4 Flexible Connections

The inlets and outlets of air supply units and exhaust fans should be connected to ductwork with flexible, airtight connectors for noise and vibration suppression. Those in exposed locations should be constructed of materials suitable for outdoor installation.

VI.a.5 Sealing Duct Penetrations

Wherever ducts pass through walls, floors, or partitions, the spaces around the ducts should be sealed with metal, mineral wool, or other noncombustible material.

APPENDIX **VII**

Emergency Showers*

1. Pull ring should not exceed 77 in. from the floor except where handicapped are involved and then the maximum should be determined functionally. (Figure VII-1)
2. The shower head should be at least 84 in. from the floor. (Figure VII-1)
3. The shower head should be an "Emergency Deluge Shower" as manufactured by the Speakman Company, (Speakman, 1986) or its equal.
4. The horizontal distance from the center of the shower head to the pull bar should not be greater than 23 in.
5. The shower should provide at least 30 gal/min flow with the operating valve in the open position.
6. Tempered water showers should be equipped with a mixing valve with an "antiscald" feature such as manufactured by Powers, Series 420 Hydroguard. (Powers, 1986)
7. Tempered water showers should be preset at a temperature between 70 and 90°F.
8. The valve for the shower should be quick acting, such as a ball valve, and should remain open after the initial pull until manually closed.

* For additional information see the American National Standards Institute standard on emergency showers (ANSI.Z358.1, 1985).

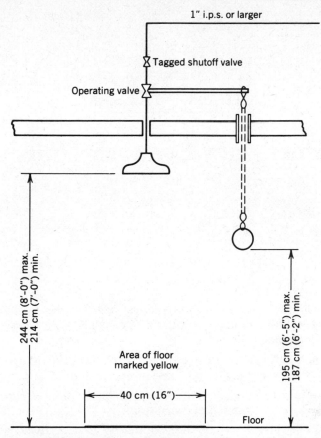

1" i.p.s. or larger

Tagged shutoff valve

Operating valve

244 cm (8'-0") max.
214 cm (7'-0") min.

195 cm (6'-5") max.
187 cm (6'-2") min.

Area of floor
marked yellow

|←——40 cm (16")——→|

Floor

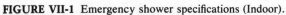

FIGURE VII-1 Emergency shower specifications (Indoor).

APPENDIX VIII

Emergency Eyewash Units

1. Eyewash units shall be approved as such by the manufacturer. Haws Drinking Fountain Co., "EMERGENCY EYEWASH" or its equivalent should be used.
2. The water supply must be potable and capable of providing 3 to 7 gal per minute depending on the type of eyewash used.
3. For each floor of the building or for each group of labs there should be at least one tempered eyewash unit.
4. Tempered water eyewash units should be equipped with a mixing valve with an "antiscald" feature such as manufactured by Powers, Series 420 Hydroguard.
5. Tempered water eyewash units should be preset to a temperature of 70 ± 5°F.

APPENDIX IX

Excess Flow Check Valves

Excess flow check valves, also known as "flow limit valves" by the Tescom Corporation (Tescom, 1987), a manufacturer of such valves, are simple devices which can provide control of gas flow for many laboratory operations. The valves are designed to shut off gas flow if a preset flow rate is exceeded. This could prevent the flow of toxic or flammable gases into an area when other conditions have resulted in failure of point-of-use control systems; i.e. a needle valve. Such control is particularly important where compressed gases are piped through a building or from one room to another.

Figure IX-1 (Tescom, 1987) shows a flow limit valve which can operate at pressures between 130 and 3000 psig and deliver flow rates from 100 to 10,000 standard cubic feet per minute. This particular unit is normally installed in the delivery line between the cylinder and process.

Figure IX-2 is a functional schematic of the valve shown in the open position to demonstrate how it functions. It is possible to fabricate these in-house utilizing one's own design.

This appendix has been included because many laboratory designers and users are unaware of their existence and use. Their use must be considered in the design stages since it may affect decisions regarding tank farms and gas piping.

FIGURE IX-1. Excess flow check valve: Commercial.

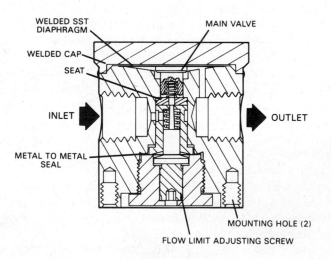

WELDED SST
DIAPHRAGM

WELDED CAP

SEAT

MAIN VALVE

INLET

OUTLET

METAL TO METAL
SEAL

MOUNTING HOLE (2)

FLOW LIMIT ADJUSTING SCREW

FUNCTIONAL SCHEMATIC
(valve shown open)

FIGURE IX-2. Functional schematic of excess flow check valve

APPENDIX X

Signs

Signs for laboratories are used for many purposes. Among them are:

Identifying exits and safety equipment and procedures.
Identifying electrical, piping, plumbing, and other facility-type equipment.
Identifying hazardous materials, equipment, and special conditions.

The first two have been well defined and standard methods exist for indicating exits, pipe content, and electrical runs and panels. As for hazardous materials, equipment, and special conditions, much is covered by agency requirements for radioactive materials, lasers, biological hazards, and the like. The problem that resulted in the development of the sign system shown below relates to hazardous chemicals (e.g., flammables, toxics, and explosives).

Two surveys were made of laboratories across the country, including university, government, industrial or commercial, and nonprofit establishments to determine what building and/or lab room sign system was in use to assist fire department or emergency response personnel—the primary purpose of such a system. The results showed that no uniform system seemed to have worked well, including NFPA 704M (NFPA, 1985). However, many useful ideas were gleaned from the study (Gatwood, 1985).

The following policy was derived from the study. The ground rules are included to help understand the sytem. They are:

- Those using NFPA 704 were experiencing difficulties including training, maintenance, quantity decision, improper information, and lack of information.
- Fire department and emergency response personnel can be expected to know that laboratories generally contain some flammable liquids, toxic chemicals, and compressed gases, and various pieces of equipment.
- Fire department and emergency response personnel may already have some familiarity with laboratories through in-service inspections, site visits to grant annual permits/licenses, plan review, and special reviews by owner/occupant.
- Signs on doors alone may be insufficient to help a fire fighter, particularly when fire and/or smoke render them unreadable.
- The sign system needs to be as simple and yet as effective as possible.
- Fire department and emergency response personnel should not have to learn, understand, or commit to memory unique systems—the information should be self-evident.
- The need for frequent changes to signs should be eliminated, so the signs will always be accurate, given the administrative difficulty and history of maintaining signs.
- Experiments move from room to room within a research group.
- Other jurisdictional agencies require signs that may already be present and should be accepted by this regulation, for example, biohazards, radiation, strong magnetic fields, lasers, UV, explosives, and high voltage.
- Emergency response–fire department personnel can be expected to enter each building at a prescribed location, giving them access to annunciator panel and building information.

X.a GENERAL POLICY

At the *Primary Fire Department and Emergency Response Personnel Entrance* to each building containing a laboratory(ies) there should be a sign (Type I, Figure X-1) indicating the type of laboratory(ies), the common hazards expected to be encountered, and notation about special hazards and their location within the building. At this building location, if appropriate, there may be supplemental information such as building dia-

```
                    BUILDING NAME _____

                    BUILDING USE _____

                    ZONE (IF APPLICABLE)_____

          THIS BUILDING/ZONE CONTAINS WORKING QUANTITIES OF THE FOLLOWING:


                                  BEING USED      "SPECIAL CONDITION IN ROOM NO."

  FLAMMABLE LIQUIDS

  WATER REACTIVES

  HAZARDOUS BIOLOGICAL AGENTS

  HIGHLY TOXIC CHEMICALS

  COMPRESSED GAS

  EXPLOSIVES

  HIGH VOLTAGE

  LASERS

  STRONG MAGNETIC FIELDS

  RADIOACTIVES

  RADIATION

  MICROWAVE RADIATION

  OTHER


        DATE POSTED_____      _(Name)__FIRE DEPARTMENT
```

FIGURE X-1. Type 1 sign at building entrance.

grams, zone diagrams, voice alarms, and protection system(s) information.

A copy of this Type I sign, where appropriate, should be posted at each zone entrance.

A second sign (Type II, Figure X-2), should be posted on the room door of any room containing a special condition or hazard, as noted on the sign at the entrance to the building or zone.

What constitutes a special hazard or condition can be defined in general terms, such as an "appreciable quantity of water reactive metals in which a fire should not be fought with water," or "an area of a building where persons might be expected to be napping." But the specific situations would be determined on an individual basis by management and the fire department. Questions of quantity should of course, be a part of that

FIGURE X-2. Type 2 sign for doors of rooms containing special hazards.

procedure, and signs indicating those special cases would, as mentioned, be located on the room door, at the zone entrance and at the building's primary fire department and emergency personnel entrance.

Before adoption of this system, communication with the local authority having jurisdiction, such as the local fire department, should be initiated. The sign formats shown may be modified to adapt to a set of specific needs.

Matrix

This appendix has been prepared as a summary reference for key safety and health issues that should be addressed in the common type laboratories. It can be used as a checklist for laboratory design or to refer the reader to the particular section in the book for a further discussion of the item. For each item we have provided guidance as to its applicability in each type laboratory. Those where RC is indicated are the items we believe are applicable to the design and in many cases are probably required by regulation. They require careful consideration and rejection of their application to a particular design should be made at the highest level of authority on the project. Some items may not be required by regulation but we believe represent good safety practice and are indicated by PF. For example fire suppression systems are preferred in a pilot plant but may not be feasible. The applicability of some items will require a special evaluation of the particular design and needs of the laboratory and are indicated by SE. For example, the use of emergency showers in a clean room will depend on the types of chemicals used in the room. There are also some items that are definitely not recommended nor applicable to a particular laboratory and they are indicated by NR. In some cases NR is used to discourage the use of certain materials in a type laboratory. For example, flammable liquid storage is not recommended in a physics lab in order to discourage the use of flammable liquids in this type of lab.

If the reader has any doubts about the particular applicability of an item, they should refer to the section in the text for the rationale of the selection.

PARAGRAPH NUMBER	BUILDING CONSIDERATIONS	CHAPTERS														
		3	4	5	6	7	8	9	10	11	12	13	14	15	16	17
1.2.2.4	Evaluate Distribution of Mechanical Equipment and Services	RC	RC	RC	RC	RC	RC	RC	RC	RC	RC	RC	RC	RC	RC	RC
1.3.1	Directional Airflow in Laboratories	RC	RC	RC	RC	RC	RC	RC	RC	RC	RC	RC	RC	RC	RC	RC
1.3.4	Supply Air Systems	RC	RC	RC	RC	RC	RC	RC	RC	RC	RC	RC	RC	RC	RC	RC
1.3.4.2	Velocity and Quantity	RC	RC	RC	RC	RC	RC	RC	RC	RC	RC	RC	RC	RC	RC	RC
1.3.4.3	Evaluate Location of Intake to Building	RC	RC	RC	RC	RC	RC	RC	RC	RC	RC	RC	RC	RC	RC	RC
2.3.3.1.1	Location of Supply Air to Room	RC	RC	RC	RC	RC	RC	RC	RC	RC	RC	RC	RC	RC	RC	RC
1.3.8	HVAC Controls/Alarms	PF	PF	RC	PF	PF	PF	PF	PF	RC	RC	SE	SE	RC	RC	PF
1.3.8.1	Temperature Control	RC	RC	RC	RC	RC	RC	RC	RC	RC	RC	RC	RC	RC	RC	RC
1.3.8.1	Humidity	SE	SE	SE	SE	SE	SE	SE	SE	SE	RC	SE	SE	RC	RC	SE

PARAGRAPH NUMBER	BUILDING CONSIDERATIONS	CHAPTERS														
		3	4	5	6	7	8	9	10	11	12	13	14	15	16	17
1.4.1	Emergency Power Source	RC	RC	RC	RC	RC	RC	RC	RC	RC	RC	RC	RC	RC	RC	RC
1.4.4.1	Fire Detection System	RC	RC	RC	RC	RC	RC	RC	RC	RC	RC	RC	RC	RC	RC	RC
1.4.4.2	Fire Suppression System	PF	SE	PF	PF	SE	SE	PF	PF	PF	PF	RC	RC	RC	RC	SE
1.4.5	Fire Alarm System	RC	RC	RC	RC	RC	RC	RC	RC	RC	RC	RC	RC	RC	RC	RC
1.5.1	Lighting	RC	RC	RC	RC	RC	SE	SE	RC	RC	RC	RC	PF	PF	PF	RC
1.5.3	Plumbing	RC	RC	RC	RC	RC	RC	RC	RC	RC	RC	RC	RC	RC	RC	RC
2.2.1	Egress	RC	RC	RC	RC	RC	SE	RC	RC	RC	RC	RC	RC	RC	RC	RC
2.2.2	Furniture Location	RC	RC	RC	RC	RC	RC	RC	RC	RC	RC	RC	RC	RC	RC	RC
2.2.4	Equipment Location	SE	SE	SE	SE	SE	SE	SE	SE	SE	SE	SE	SE	SE	SE	SE

LEGEND:

RC = Recommended for Considered Application

PF = Preferred for Considered Application

SE = Special Evaluation Needed

NR = Not Recommended or Not Applicable

CHAPTERS:

3) General Chemistry
4) Analytical Chemistry
5) High Toxicity
6) Pilot Plant
7) Physics
8) Clean Room
9) Controlled Environment
10) High Pressure
11) Radiation
12) Biological
13) Clinical
14) Teaching
15) Gross Anatomy
16) Pathology
17) Team Research

PARAGRAPH NUMBER	BUILDING CONSIDERATIONS	CHAPTERS														
		3	4	5	6	7	8	9	10	11	12	13	14	15	16	17
2.2.5	Handicapped Access	RC	RC	RC	RC	RC	RC	RC	RC	RC	RC	RC	RC	RC	RC	RC
2.3.4.3	Laboratory Chemical Fume Hood	RC	RC	RC	SE	RC	SE	SE	SE	RC	SE	RC	RC	RC	RC	RC
2.3.4.3.3	Perchloric Acid Hood	SE	SE	SE	SE	SE	NR	NR	NR	SE	SE	SE	NR	NR	SE	SE
2.3.4.3.4 & 12.3.2	Biological Safety Cabinet	NR	NR	SE	NR	NR	NR	NR	NR	NR	RC	RC	NR	NR	RC	NR
2.3.4.3.5	Local Exhaust Ventilation	SE	PF	RC	RC	SE	SE	SE	SE	PF	SE	RC	SE	SE	RC	PF
2.4.1.1	Emergency Gas Shut-Off	RC	RC	RC	RC	RC	RC	RC	RC	RC	RC	RC	RC	RC	RC	RC
2.4.1.2	Ground Fault Circuit Interruptors	RC	RC	RC	RC	RC	RC	RC	RC	RC	RC	RC	RC	RC	RC	RC
2.4.1.3	Master Electrical Disconnect Switch	RC	RC	RC	RC	RC	RC	RC	RC	RC	RC	RC	RC	RC	RC	RC
2.4.1.4	Emergency Showers	RC	RC	RC	RC	SE	SE	SE	RC	RC	RC	RC	RC	RC	RC	RC

PARAGRAPH NUMBER	BUILDING CONSIDERATIONS	3	4	5	6	7	8	9	10	11	12	13	14	15	16	17
2.4.1.5	Emergency Eye Wash	RC	RC	RC	RC	SE	SE	SE	RC	RC	RC	RC	RC	RC	RC	RC
2.4.1.6	Chemical Spill Control	PF	PF	RC	PF	PF	PF	PF	PF	PF	PF	RC	RC	RC	RC	PF
2.4.2	Construction Methods and Materials	RC	RC	SE	RC	RC	RC	SE	SE	RC	RC	RC	RC	RC	RC	RC
2.4.4	Experiment Alarm Systems	PF	PF	PF	PF	PF	PF	PF	PF	PF	PF	NR	SE	SE	NR	PF
2.4.5 & 1.4.6	Hazardous Chemical Disposal	RC	RC	RC	RC	RC	RC	RC	RC	RC	RC	RC	RC	RC	RC	RC
2.4.5.2	Chemical Waste Treatment	RC	RC	RC	RC	RC	RC	RC	RC	RC	RC	RC	RC	RC	RC	RC
2.4.6.3	Flammable Liquid Storage	RC	RC	RC	PF	NR	NR	NR	PF	PF	PF	RC	RC	RC	RC	PF
2.4.6.4	Special Hazard Chemicals	PF	PF	PF	PF	RC	PF	PF	PF	PF	PF	NR	NR	NR	NR	PF
2.4.7	Compressed Gas Cylinders	PF	PF	PF	PF	PF	PF	PF	PF	PF	PF	PF	PF	PF	PF	PF

LEGEND:

RC = Recommended for Considered Application
PF = Preferred for Considered Application
SE = Special Evaluation Needed
NR = Not Recommended or Not Applicable

CHAPTERS:
3) General Chemistry
4) Analytical Chemistry
5) High Toxicity
6) Pilot Plant
7) Physics
8) Clean Room
9) Controlled Environment
10) High Pressure
11) Radiation
12) Biological
13) Clinical
14) Teaching
15) Gross Anatomy
16) Pathology
17) Team Research

PARAGRAPH NUMBER	BUILDING CONSIDERATIONS	CHAPTERS														
		3	4	5	6	7	8	9	10	11	12	13	14	15	16	17
2.4.8	Emergency Cabinet	PF	PF	PF	PF	PF	PF	PF	PF	PF	PF	RC	RC	RC	RC	PF
2.5	Special Services	SE	SE	SE	SE	SE	SE	SE	SE	SE	SE	SE	SE	SE	SE	SE
2.5	Special Services	RC	RC	RC	RC	RC	RC	RC	RC	RC	RC	RC	RC	RC	RC	RC
5.2.2	Change Room	NR	NR	RC	NR	NR	RC	NR	NR	SE	NR	NR	NR	SE	SE	NR
5.2.3	Work Surfaces	PF	PF	RC	PF	PF	PF	PF	PF	RC	RC	PF	PF	PF	PF	PF
5.2.4	Floors and Walls	PF	PF	RC	PF	PF	PF	PF	PF	RC	RC	PF	PF	PF	PF	PF
5.2.5	Handwashing Facilities	PF	PF	RC	PF	PF	PF	PF	PF	RC	RC	RC	RC	RC	RC	PF
5.2.6	Access Restrictions	PF	PF	RC	PF	PF	RC	PF	RC	RC	RC	PF	PF	RC	RC	PF
5.3.3	Glove Box	NR	NR	RC	SE	NR	SE	NR	NR	RC	NR	NR	NR	NR	NR	NR

272

PARAGRAPH NUMBER	BUILDING CONSIDERATIONS	CHAPTERS														
		3	4	5	6	7	8	9	10	11	12	13	14	15	16	17
5.3.6	Filtration of Exhaust Air	SE	SE	RC	SE	SE	SE	SE	SE	SE	SE	SE	SE	SE	SE	SE
5.4.2	Protection of Laboratory Vacuum System	SE	SE	RC	SE	SE	SE	SE	SE	SE	SE	SE	SE	SE	SE	SE
11.3.2	Radioisotope Hood	NR	SE	SE	SE	SE	NR	NR	SE	RC	SE	SE	NR	NR	SE	SE

LEGEND: RC = Recommended for Considered Application

PF = Preferred for Considered Application

SE = Special Evaluation Needed

NR = Not Recommended or Not Applicable

CHAPTERS:
3) General Chemistry
4) Analytical Chemistry
5) High Toxicity
6) Pilot Plant
7) Physics
8) Clean Room
9) Controlled Environment
10) High Pressure
11) Radiation
12) Biological
13) Clinical
14) Teaching
15) Gross Anatomy
16) Pathology
17) Team Research

273

References

ACGIH (1984)   American Conference of Governmental Industrial Hygienists, 1984, *Industrial Ventilation, A Manual of Recommended Practices,* 18th ed., Committee on Industrial Ventilation, Lansing, MI, 1984.

AMCA (1974)   *Laboratory Methods of Testing Fans for Rating,* AMCA Std. 210-74, Arlington Heights, IL.

ANSI (1980)   *Standard for Testing Nuclear Air-Cleaning Systems,* ANSI/ASME N510, American Society of Mechanical Engineers, New York.

ANSI-ASHRAE (1986)   *Ventilation for Acceptable Indoor Air Quality,* ANSI/ASHRAE Standard 62-1981R, American Society of Heating, Refrigerating and Air Conditioning Engineers, Atlanta, GA.

ASHRAE (1925)   American Society of Heating, Refrigerating and Air Conditioning Engineers, Atlanta, GA *Guide,* 1925.

ASHRAE (1977)   *Fundamentals,* 1977.

ASHRAE (1981)   *Fundamentals,* 1981.

ASHRAE (1982)   *Applications,* 1982.

ASHRAE (1983)   *Equipment,* 1983.

ASHRAE (1984)   *Systems,* 1984.

ASHRAE (1985)   *Fundamentals,* 1985.

ASHRAE (1986)   *Refrigeration,* 1986.

ASME (1983)   American Society of Mechanical Engineers, 1983 ASME Unfired Pressure Vessel Code, Section VIII, ASME, Fairfield, NJ

ASSE (1977)   Weaver, A., and Britt, K., "Criteria for Effective Eyewashes and Safety Showers," *Professional Safety Magazine,* pp. 38–54,

June 1977, American Society of Safety Engineers, Des Plaines, IL

BOCA (1985)   The Building Officials and Code Administrators International, Inc., Chicago, IL, 1985.

BSI (1979)   British Standards Institution, *Draft Standard for Safety Requirements for Fume Cupboards, Performance Testing, Recommendations on Installation and Use,* BSI Document 79/52625 DC, London, United Kingdom, August, 1979.

Buffalo Forge (1983)   Jorgensen, R., Ed., *Fan Engineering,* 8th ed. Buffalo Forge, Buffalo, NY, 1983.

Burgess (1985)   Burgess, W. A. and Forster, F., "The Evaluation of gas cabinets used in the semiconductor industry," in *Ventilation '85,* H. D. Goodfellow, Ed., Elsevier, Amsterdam, 1986.

Caplan (1978)   Caplan, K. J. and Knutson, G. W., "Laboratory fume hoods: A performance test," *ASHRAE Transactions,* **84,** Part 1, 1978.

CDC (1984)   Centers for Disease Control and National Institutes of Health, *Biosafety in Microbiological and Biomedical Laboratories,* HHS Publication No. (CDC) 84-8395, U.S. Government Printing Office, Washington, DC.

Chamberlin (1978)   Chamberlin, R., and Leahy, J., *Laboratory Fume Hood Standards,* EPA Report 68-01-4661, 1978.

Chamberlin (1979)   Chamberlin, R., *Laboratory Fume Hood Specifications and Performance Testing Requirements,* Environmental Medical Service, Massachusetts Institute of Technology, Cambridge, 1979.

Chamberlin (1982)   Chamberlin, R. I. and Leahy, J. E., *Development of Quantitative Containment Performance Tests for Laboratory Fume Hoods,* EPA Contract No. 68-01-6197, June, 1982.

CMR (1979)   Codes of Massachusetts Regulations, "Cross Connections—Drinking Water," 310 CMR 22.22, Massachusetts Department of Environmental Quality Engineering, Boston, MA, Sept. 20, 1979.

DeRoss (1979)   DeRoss, Roger, et al., *Hospital Ventilation Standards and Energy Conservation: A Summary of Literature with Conclusions and Recommendations,* FY79 Final Report, University of Minnestoa, 1979.

DiBerardinis (1983)   DiBerardinis, L.J., et al., "Storage Cabinet for Volatile Toxic Chemicals," *American Industrial Hygiene Association Journal* **44** (8), 583–588 (1983).

Fawcett (1984)   Fawcett, H., *Hazardous and Toxic Materials,* Wiley, New York, 1984.

First (1977)   First, M. W., *Control of Systems, Process and Operations,* 3rd ed., Vol. 4, *Air Pollution,* A.C. Stern, Ed., Chap. 1, Academic Press, New York, 1977.

FM (1982)   Factory Mutual, *Factory Mutual Approval Guide 1982,* FM, Norwood, MA, 1982.

Fuller (1979)   Fuller, F. H. and Etchells, A. W., "The rating of laboratory hood performance," *ASHRAE Journal,* 49-53 (October, 1979).

Gatwood (1985)   Gatwood, G. T. and Fresina, J., Questionnaire survey on labora-
                        tory hazards warning systems conducted in 1983 and reported
                        to the Cambridge, MA Fire Department in October, 1985.

GSA (1979)   *Public Buildings Service Guide Specifications, Airflow Control
                        Systems,* General Service Administration PBS (PCD): 15980,
                        Washington, DC, 1979.

Inglis (1980)   Inglis, J.K., Pergamon Press, Oxford, 1980 (Maxwell House,
                        Fairview Park, Elmsford, NY).

Ivany (1986)   Ivany, R., DiBerardinis, L., and First, M. W., "Laboratory fume
                        hoods: A field quantitative performance test," presented at
                        American Industrial Hygiene Conference, Dallas, TX, May,
                        1986.

JCAH (1987)   Joint Commission on Accreditation of Hospitals, *Accreditation
                        Manual for Hospitals,* 1987, Chicago, IL.

Jones (1966)   Jones, A. R., and Hopkins, D. J., "Controlling Health Hazards in
                        Pilot Plant Operations," *Chemical Engineering Progress* **62**
                        (12), 59–67 (1966).

Mass (1982)   Massachusetts Register, Architectual Barriers Board, *1982 Rules
                        and Regulations.*

Mikell (1981)   Mikell, W. G. and Hobbs, L. R., "Laboratory hood studies,"
                        Journal of Chemical Education, **58**(5), A165-A170 (1981).

McQuiston (1977)   McQuiston, F. C. and Parker, J. D., *Heating, Ventilating and Air
                        Conditioning Analysis and Design,* Wiley, New York, 1977.

NAS (1976)   National Academy of Sciences, *ILAR News,* Vol. XIX, No. 4,
                        Summer 1976, NAS, Long-Term Holding of Laboratory Ro-
                        dents, Institute of Laboratory Resources.

NFPA (1985)   National Fire Protection Association, Batterymarch Park,
                        Quincy, MA. Individual items are:
                        NFPA 10   Portable Fire Extinguishers
                        NFPA 12   Carbon Dioxide Systems
                        NFPA 12A   Halon 1301 Systems
                        NFPA 13   Sprinkler Systems
                        NFPA 30   Flammable and Combustible Liquid Code
                        NFPA 45   Fire Protection for Laboratories Using Chemicals
                        NFPA 56C   Laboratories in Health Related Institutions
                        NFPA 68   Explosion Venting Guide
                        NFPA 70   National Electric Code
                        NFPA 72A   Local Protection Signaling Systems
                        NFPA 72E   Automatic Fire Detectors
                        NFPA 90A   Installation of Air Conditioning and Ventilation Sys-
                                   tems
                        NFPA 91   Blower and Exhaust Systems for Dust, Stock and
                                   Vapor Removing or Conveying
                        NFPA 101   Life Safety Code
                        NFPA 704   Identification of the Fire Hazards of Materials

NIH (1981)   *NIH Guidelines for the Laboratory Use of Chemical Carcino-
                        gens,* U.S. Department of Health and Human Services May
                        1981.

NIH (1985) *Guide for the Care and Use of Laboratory Animals,* Publication No. 85-23, National Institutes of Health, Bethesda, MD, 1985.

NRC (1981) U.S. Nuclear Regulatory Commission, *Radiation Surveys in Medical Institutions,* Regulatory Guide 8.23, Government Printing Office, Washington, D.C., January, 1981.

NSF (1983) National Sanitation Foundation, Standard No. 49 for Class II (Laminar Flow) Biohazard Cabinetry, NSF, Ann Arbor, MI, May 1983.

Nuffield (1961) Nuffield Foundation, *The Design of Laboratory Buildings,* Oxford University Press, Oxford, 1961.

OSHA (1978) Occupational Safety and Health Administration, *General Industry,*
OSHA Safety and Health Standards (29 CFR 1910)
OSHA 2206, Revised November 7, 1978,

Pipitone (1984) Pipitone, David A., Ed., *Safe Storage of Laboratory Chemicals,* Wiley, New York, 1984.

Powers (1986) Powers Process Controls, 3400 Oakton St., Skokie, IL.

SAMA (1975) Scientific Apparatus Makers Association, *Standard for Laboratory Fume Hoods, Laboratory Equipment II,* Washington, D. C., 1975.

SMACNA (1984) Sheet Metal and Air Conditioning Contractors National Assoc. Inc., *Thermoplastic Duct (PVC) Construction Manual,* 8224 Old Court House Road, Tyson Corner, Vienna, VA.

SMACNA (1985) Sheet Metal and Air Conditioning Contractors National Assoc. Inc., *HVAC Duct Construction Standards: Metal and Flexible,* 8224 Old Court House Road, Tyson Corner, Vienna, VA.

Speakman (1986) Speakman Company, Safety Equipment Division, Wilmington, DE.

Stoeker (1958) Stoeker, W. F., *Refrigeration and Air Conditioning,* McGraw-Hill, New York, 1958.

Tescom (1987) Tescom Corporation, Elk River, MN.

Thuman (1977) Thuman, A., *Plant Engineers and Managers Guide to Energy Conservation,* Van Nostrand, New York, 1977.

UL (1984) Underwriters Laboratory, *Factory Made Air Ducts and Connectors,* Standard No. 181, Underwriters Laboratories, Inc., Northbrook, IL, 1984.

Walters (1980) Walters, Douglas B., Ed., *Safe Handling of Chemical Carcinogens, Mutagens, Teratogens and Highly Toxic Substances,* Vol. 1, Ann Arbor Science, Ann Arbor, MI, 1980.

Yaglou (1925) Yaglou, C. P., "Comfort Zones for Men at Rest and Stripped to the Waist," *Transaction of the American Society of Heating and Ventilating Engineers* **33,** 165 (1925).

Index

Acceptance, *see* Performance and final
acceptance
Access:
  for handicapped, 26, 58, 63, 123
  restrictions, 98, 101, 113, 123, 133, 139,
    182
Air:
  airflow, directional, 100, 117, 142
  balance, 197
  cleaning:
    exhaust, 74, 99, 134, 143, 184
    glove box, 242
    supply, 35, 117, 139
      standards, 37, 230
  discharge, 37, 142
  exchange rates, 42
  exhaust, 65, 66, 71, 74, 93, 99, 100, 103,
    120, 134, 147, 153, 184
  intakes, 37
  supply, 39, 67, 142, 147
    standards, 39, 230
Aisles, 59, 62, 123
Alarms:
  experimental equipment, 46, 81
  fire, 46, 108, 119, 149
  other, 46, 118, 124, 143, 185
Analytical chemistry laboratory, 92
  equipment and materials, 92
    hazardous materials, 93

  layout, 93
  local exhaust system, 93
  perchloric acid use, 94
  work activities, 92
Animal research laboratory, 175
  HVAC, 182
    alarms, 185
    controls, 184
    filtration, 182
  illumination, 186
  layout, 179
  materials of construction, 181
  pest control, 177

Benches, *see* Furniture
Bidding procedures, 189
  bidding document, 189
Biological safety cabinets, 71, 138, 143,
    147
  classes, 242, 244
  standards, 245
  UV lamps, 247
Biosafety laboratory, 136
  alarms, 143
  codes, NSF No.49, 138
  controls, 143
  decontamination, 142–144
  equipment and materials, 138
  filtration, 143

Biosafety (*Continued*)
  layout, 139, 140
  materials of construction:
    floors, 139
    walls, 139
  signs, 141–143
  ventilation, 142
Biosafety levels, 136, 137
Building:
  area:
    gross, 24
    net assignable, 14, 24
    net usable, 24
  chase, location:
    between modules, 32
    central, 32
    exterior wall, 31
    interior wall, 31
  codes, 26
  enclosure, 28
  laboratory module, 26
    length, 28, 32
    vertical, 31
    width, 27
  laboratory unit, 28
    construction, 28
    fire protection, 28
    hazard classification, 29
  layout, 12
  mechanical services, 28
  program, 12
    occupancy, 12
    outline checklist, 12, 22
    planning, 25
    space, 14, 21
  room types, 24
  site:
    air discharge, 37
    air intake, 37
    regulations, 33
  spatial organization, 25
  structure, 33
  water:
    drinking water, 53
      backflow preventers, 53
      codes, 53
      pressure, 53

Carcinogens, *see* High-toxicity laboratory
Check valves, excess flow, 76, 261

Chemical fume hood, *see* Hoods
Chemical stockroom, *see* Storage,
  chemicals
Clean room laboratory, 112
  access restrictions, 113
  classifications, 112
  cleanliness, personnel, 113
  HVAC, 117
    alarms, 118
    fans, auxiliary, 114
    filtration, 117
    fire, 119
    humidity control, 118
    pressure balance, 113, 117
    static pressure controllers, 117
    temperature control, 119
    ventilation, 117
  layout, 114, 115, 118
    air-blast chamber, 114
    decontamination, 115
    egress routes, 114
    interlocking doors, 116
    robing room, 114
    traffic flow, 115
    vestibules, 114
  lighting, 119
  materials of construction:
    ceilings, 113
    floors, 113
    walls, 113
  work activities, 112
Clinical laboratory, 145
  codes, NFPA, 149
  egress, 148
  equipment and materials, 146
  fire suppression, 149
  fume hood, 147
  HVAC, 146
    exhaust, local, 147
    supply air, 147
    temperature control, 148
  layout, 146
Codes, 7, 26, 43, 57, 63, 138, 149, 253
  references, 273–276
Compressed gases:
  piping, 48
  storage, 48
Controlled environment room, 120
  equipment and materials, 120
  exclusions, 121

humidity control, 124
layout, 121, 122
  egress, 122
  furniture, 123
  handicapped access, 123
  size, 121
lighting requirements, 125
  equipment, 125
materials of construction, 125
pretesting, 121
respirators, 125
temperature control, 124
work activities, 120
Controls, experimental equipment, 44, 81
  emergency alarm and controls, 124

Dampers, 75
  trimming, 40
Decontamination, 97, 115, 141–144, 168,
  169, 186
Disposal, chemicals, 82
Ducts:
  codes, 253
  design specifications, 255
  exhaust, 40, 74, 253
    access doors, 256
    construction, 74
    dampers, 75, 256
    fire stops, 256
    flexible, 75, 256
    leakage, 75
    sealing penetrations, 257
    self-balancing, 75
    testing, 75
  materials of construction, 40, 254
  standards, 40
  supply, 40
    leakage, 40

Egress, 56, 173. See also Aisles
  codes, 56, 64
  direction of door swing, 56
  doors:
    dimensions, 58
    glass, 56
    handles, 57
    number, 56
    protrusion, 56
    swing, 56

for handicapped, 58, 63
number of exits, 56
Electrical:
  emergency considerations, 42
  emergency generators, 43, 201
  ground fault interrupters, 77
  grounding, 78, 111
  master disconnect, 78
  outlets, 81
Emergency:
  control, 80
    chemical spills, 80
    emergency cabinet, 80
  power, 42, 43
Energy conservation, 202
  auxiliary air hoods, 207
  heat recovery system, 208
  humidity control, 212
  lighting, 212
  limit exhaust from hoods, 205
  reduced operating time, 203
  thermal insulation, 212
Entry, see Egress
Exhaust, see Air
Eyewash, 79, 108, 152, 154, 174
  plumbed, 79
  portable, 85
  specifications, 259

Face velocity, see Hoods
Fans, 227
  centrifugal, 229
  exhaust, 38, 72
    explosion proof, 73
    materials of construction, 38, 72
    raincaps, 39
  motors, 74
  specifications, 229
  terminology, 227–228
  vane-axial, 229
Filters, selection, 230. See also Air,
  cleaning
Fire:
  detection, 44, 108
    heat sensitive, 44
    ionization, 44
    LED-operated, 44
  suppression:
    fixed automated systems, 45
    hand-portable extinguishers, 45

Flammable liquid, *see* Storage
Flow limit valve, 76, 261
Fume cupboard, *see* Hoods, bypass
Fume hoods, *see* Hoods
Furniture:
  benches, 60
    work surfaces, 62
  desks, 62
  hoods, 62
  location, 60–63

Gas, fuel, 76
  check valve, 76
    specifications, 260
  shutoff, 49, 76
Gases, compressed:
  storage, 84
General chemistry laboratory, 89
  equipment and materials, 89
  exclusions, 90
  layout, 90, 91
  work activities, 89
Glove boxes, 68, 99, 133, 241, 242
  air cleaning, 242
  materials of construction, 242
Gross anatomy laboratory, 156
  cold storage, 159
  dissection, 157
  equipment and materials, 157
  exclusion, 159
  HVAC, 161
  layout, 159
  morgue, 156
  research, 157
Ground fault interrupter, 78

Handicapped persons:
  access, 58, 63, 123
  codes, 26
Hazardous waste:
  chemicals, 82
  storage room, 47
Heating, Ventilating, Air Conditioning, *see*
    HVAC
High-pressure laboratory, 126
  containment, 127
    berms, 128
    blowout panels, 128
  equipment and materials, 127
  fire extinguishers, 129

gas cylinders, 129
layout, 127
location, 128
materials of construction, 127
pressure vessels, 128
ventilation:
  equipment:
    ducts, 129
    fans, 129
    vents, 129
  rates, 129
  vents, 127, 28
work activities, 126
High-toxicity laboratory, 95
  access restrictions, 96, 98
    signs, 98
  equipment and materials, 96
  exclusions, 96
  exhaust air, filtration, 99, 100
  floors, 97
    penetrations, 97
  glove box, 99
  hood face velocity, 98
  layout, 96
    change room, 97
    decontamination room, 97
    hand washing, 98
    shower room, 97
    traffic flow, 97
  local exhaust, 99
  ventilation:
    pressure gradients, 100
    storage facilities, 99
  walls, 97
    penetrations, 97
  work:
    activities, 95
    surfaces, 97
Hoods, 67, 231
  auxiliary air, 69, 238
    specification, 240
  biological safety, 71, 242
    classes, 242, 243
    types, 243
  bypass, 232
    air foils, 233, 235
    construction, 233–238
    sash, 232
    specification, purchase, 234–238
  cabinets, support, 237

canopy, 248
capture, 248
chemical storage, 237
electrical, 237
face velocity, 98, 234, 236
horizontal sliding sash, 70, 238
manifolded, 74
perchloric acid, 70, 240
performance characteristics, 68
performance tests, 250, 251
radioactive, 241
  face velocity, 241
slot, 148
Humidity control, see HVAC
HVAC, 6, 34, 36, 215. See also Air;
  Laboratory, types
air conditioning, 215
  central system, 217
  distribution system, 219–221
  modular system, 217
air exchange rate, 41
balance, 197
building, 34
  air exchange rates, 41
  controls:
    humidity, 41, 118, 212
    temperature, 42
  exhaust, 35
    discharge, 37
  materials of construction, 43
    codes, 43
    fire resistive, 43
  pressure relationships, 34, 36
  supply air, 35
    cleaning, 39–40
    intake, 37
    velocity, 37
    volume, 36
  support services, 54
chiller, see HVAC, evaporators
comfort, 222, 226
  standards, 224–226
constant volume, 36
control systems, 41, 64
cooling systems, 217
  compressors, 217, 218
  condensers, 218
  evaporators, 219
directional airflow, 100
effective temperature, 222

light work, 224
resting, 223, 225
exhaust air, 66
  velocity, 67
  ventilation rate, 66
heat gains and losses, 216
  cooling load, 216
  heating load, 216
  insolation, 216
humidity control, 41, 65, 212
modulating volume, 36
noise, 76
pressure, 64
  hazardous materials, 65
  reserves, 64
supply air, 65. See also Air
  location, 65
  recirculation, 66
  velocity, 65
systems, 35
  branched, 35
  dedicated, 35
temperature control, 41, 64

Illumination, 50
animal laboratory, 186
Information sources, 7

Laboratory:
  equipment, safe locations, 84
  furniture, 59
  hoods, see Hoods
  HVAC, see HVAC
  matrix, 268
  module, 26
  types:
    analytical chemistry, 92
    animal research, 175
    biosafety, 136
    chemical engineering, 101
    clean room, 112
    clinical, 145
    controlled environment, 120
    dissection, 156
    general chemistry, 89
    gross anatomy, 156
    high-pressure, 126
    high-toxicity, 95
    pathology, 164
    physics, 105

Laboratory (*Continued*)
  pilot plant, 101
  radiation, 130
  teaching, 150
  team research, 170
  unit, 24
Layout, laboratory, 55. *See also*
  Laboratory, types
Lighting levels, 50

Materials of construction, 26, 80, 125
  fire resistive, 80
Matrix, 266
Morgue, 156, 164
Mutagens, *see* High-toxicity laboratory

Noise suppression, 77, 158

Pathology laboratory, 164
  clinical morgue, 164
  cold storage, 165
  equipment and materials, 165
  HVAC, 167
  layout, 167
  sample preparation room, 165
Perchloric acid hood, 70, 240
Performance and final acceptance, 193
  check lists, 194
    alarm systems, 200
    construction, 196
    design, 194
    emergency electrical system, 201
    emergency showers, 201
    eyewash, 201
    HVAC, 197
      air balancing, 197
      duct work testing, 200
      fume hoods, 198
    pre-occupancy, 194
Pest control, 177
Physics laboratory, 105
  equipment and materials, 106
  eyewash, 108
  fire:
    alarms, 108
    detection, 108
    extinguishers, 109, 110
    suppression, 109
  hazards:
    electrical, 110

grounding, 111
  materials, 110
layout, 106, 107
  egress routes, 107
  lasers, 107
water requirements, 111
work activities, 105
Pilot plant, 101
  access restrictions, 101
  equipment and materials, 102
  isolation, 101
  layout, 102
  lighting, 104
  respirators, 104
  spill containment, 103
  training, personnel, 102
  ventilation
    general, 103
    local exhaust, 103
    needs, 102
  work activities, 101
Planning, 24
Plenums, *see* Ducts
Plumbing, 51
  dilution tanks, 51
  drains, 111
  drinking water protection, 53
  pressure, 53
  sinks, 51
  wastes, 51
  water supplies, 111
Program, building, 12

Radiation laboratory, 130
  access 131, 133
  definition, 131
  equipment and materials, 131
  exhaust ventilation, 133
    filtration, 134
      bag-in, bag-out, 134
    glove box, 133
      pressure differential, 133
    hood, isotope, 133
    local (spot) exhaust, 134
    monitoring, activity levels, 134
  layout, 131
    change rooms, 132
    hand wash facilities, 132
    showers, 132
  materials of construction:

floors, 132
  shielding, 135
  walls, 132
  work surfaces, 132
permits, 131
radioactive waste, 135
  shielding, 135
  shipping, 135
  storage, 135
signs, 133
storage, radioactive materials, 134
ventilation, 134
windows, 133
work activities, 130
Renovation projects, 87
  general principles, 3
Respirators, compressed air for, 104

Safety controls, 44, 81
Shielding, radioactive, 135
Showers:
  emergency, 78, 153
  specifications, 257
Signs, 262
  biohazard, 141
  codes, 49
  emergency response personnel, 263
  equipment, 49
  hazardous materials, 49
  safety procedure, 49
Sinks, 51
Smoke detection, see Fire, detection
Spills, containment, 103
Spot exhaust, 71
Stacks, 72
Standards, see Codes
Storage:
  chemicals, 48, 83–84
  compressed gases, 49
  flammable liquids, 83
  hazardous wastes, 48
Supply, air, see Air
Surfaces, work, 62, 97, 132

Teaching laboratory, 150
  chemical disposal, 154
  chemical storage, 154
  egress, 152
  emergency showers, 153
  equipment and materials, 151
  exclusions, 151
  exhaust, local, 153
  eyewash, 154
  fire extinguishers, 154
  HVAC, 153
  layout, 151
  preparation room, 155
Team research laboratory, 170
  egress, 173
  emergency shower, 174
  equipment and materials, 171
  exclusions, 171
  eyewash, 174
  fire extinguishers, 174
  HVAC, 173
  layout, 171
  utility distribution, 172
  zoning, 172
Temperature, see HVAC
Teratogens, see High-toxicity laboratory

Ventilation, see HVAC

Waste:
  animal, 176
  chemical, 154
  hazardous, 47
  liquid, 51
  pathology, 169
  radioactive, 135
  storage, 47
  trash, general, 47
  treatment, 82
Water:
  drinking, protection, 53
  pressure, 53
  supply, 111